State of the Art of Natural Antioxidants: Extraction, Detection and Biofunctions

State of the Art of Natural Antioxidants: Extraction, Detection and Biofunctions

Guest Editors

Mostafa Gouda
Yong He
Alaa El-Din A. Bekhit
Xiaoli Li

MDPI

Basel • Beijing • Wuhan • Barcelona • Belgrade • Novi Sad • Cluj • Manchester

Guest Editors

Mostafa Gouda
Department of Nutrition & Food Science
National Research Centre
Giza
Egypt

Yong He
College of Biosystems Engineering and Food Science
Zhejiang University
Hangzhou
China

Alaa El-Din A. Bekhit
Department of Food Science
University of Otago
Dunedin
New Zealand

Xiaoli Li
College of Biosystems Engineering and Food Science
Zhejiang University
Hangzhou
China

Editorial Office
MDPI AG
Grosspeteranlage 5
4052 Basel, Switzerland

This is a reprint of the Special Issue, published open access by the journal *Molecules* (ISSN 1420-3049), freely accessible at: https://www.mdpi.com/journal/molecules/special_issues/ultrasound_sensors.

For citation purposes, cite each article independently as indicated on the article page online and as indicated below:

Lastname, A.A.; Lastname, B.B. Article Title. *Journal Name* **Year**, *Volume Number*, Page Range.

ISBN 978-3-7258-5881-1 (Hbk)
ISBN 978-3-7258-5882-8 (PDF)
https://doi.org/10.3390/books978-3-7258-5882-8

Contents

About the Editors

Mostafa Gouda

Mostafa Gouda is a Professor (Associate/Adjunct) at the Department of Nutrition and Food Science, National Research Centre, Egypt, and College of Biosystems Engineering and Food Science, Zhejiang University, China. His research interests focus on analytical chemistry, biosensing, and advanced spectroscopic methods (including micro-Raman systems, hyperspectral imaging, and other technologies) for biosystems' advances, phytomedicine, and environmental applications. He has coordinated several international projects on bioresource valorization, microalgae-based detection, and AI-assisted quality assessment. Dr. Gouda has authored and co-authored numerous peer-reviewed articles in top-tier journals such as *Trends in Plant Science, Biosensors and Bioelectronics, Food Hydrocolloids, Engineering, Food Chemistry,* and *Plant Physiology and Biochemistry*. He serves as Associate Editor and Editorial Board Member for various international journals and has been recognized with prestigious awards, including the Best Young Researcher Award (ASGEB, 2025) and the National State Award of Egypt (2023).

Yong He

Yong He is a Distinguished Professor at the College of Biosystems Engineering and Food Science, Zhejiang University, China. His research focuses on intelligent sensing and spectroscopy for food and agricultural systems, particularly hyperspectral imaging, terahertz spectroscopy, and machine vision for quality assessment and precision agriculture. He leads multiple national and international research projects and has published extensively in *Trends in Food Science & Technology, Analytica Chimica Acta,* and *Biosensors and Bioelectronics*. Professor He is recognized among the top scholars in agricultural and biological engineering and serves on editorial boards of several high-impact journals. His pioneering work in integrating artificial intelligence with non-destructive sensing technologies has significantly advanced modern precision food analysis.

Alaa El-Din A. Bekhit

Alaa El-Din A. Bekhit is a Professor in the Department of Food Science at the University of Otago, New Zealand. His research covers meat science, functional foods, natural antioxidants, and bioactive compound extraction. He has contributed to the development of sustainable and health-promoting food products, with special emphasis on antioxidant-rich functional ingredients. Professor Bekhit has published more than 200 scientific papers and several book chapters, receiving international recognition for his work in food chemistry and technology. His collaborative projects span across continents, reflecting his leadership in global food innovation, value addition, and the valorization of by-products for human nutrition.

Xiaoli Li

Xiaoli Li is a Professor at the College of Biosystems Engineering and Food Science, Zhejiang University, China. Her research interests include spectroscopy, machine learning, and bioinformatics for the evaluation of agricultural and food products. She specializes in terahertz and hyperspectral imaging technologies for the non-destructive assessment of biomolecules and natural antioxidants. Professor Li has published widely in international journals and has co-led several key national and provincial research projects on intelligent sensing and precision quality control. Her interdisciplinary approach bridges analytical chemistry and biosystems engineering, contributing to innovative methodologies for food quality and safety evaluation.

Preface

This Reprint collects a selection of cutting-edge articles that explore the intersection of analytical techniques and chemical/biological sensing, aiming to showcase how these methods could be powerful tools for molecular detection, reaction enhancement, and sensor development. It intends to highlight the recent advances that push the frontiers of niche physical techniques into a versatile modality that can influence chemistry (e.g., sonochemistry).

Mostafa Gouda, Yong He, Alaa El-Din A. Bekhit, and Xiaoli Li
Guest Editors

molecules

MDPI

Editorial

Emerging Technologies for Detecting the Chemical Composition of Plant and Animal Tissues and Their Bioactivities: An Editorial

Mostafa Gouda [1,2,*], Yong He [1,*], Alaa El-Din Bekhit [3] and Xiaoli Li [1,*]

1 College of Biosystems Engineering and Food Science, Zhejiang University, 866 Yuhangtang Road, Hangzhou 310058, China
2 Department of Nutrition & Food Science, National Research Centre, Dokki, Giza 12422, Egypt
3 Department of Food Sciences, University of Otago, Dunedin 9054, New Zealand; aladin.bekhit@otago.ac.nz
* Correspondence: mostafa-gouda@zju.edu.cn or goudarowing@yahoo.com (M.G.); yhe@zju.edu.cn (Y.H.); xiaolili@zju.edu.cn (X.L.)

Integrating physical and chemical technologies for the characterization and modification of plants and animal tissues has been used for several decades to improve their detection potency and quality [1]. Scientists have been exploring the scientific basis and mechanism of action of the chemical constituents in biological tissues [1,2]. Additionally, special attention has been paid to investigating the different methods for maximizing the detection efficacy of their bioactivities and understanding the changes in their chemical compositions [3–6]. As an example, using ultrasound (US) in detecting the chemical composition of the biological tissues and their bioactivities has become an important emerging technologies [7,8]. Bourdeau, et al. [9] developed an acoustic method for visualizing and imaging the microbial cellular chemical composition inside mammalian hosts in vivo. In addition, these methods proved their efficiency in their application as analytical methods. For example, acoustic sensors based on quartz crystal microbalance (QCM) were used to detect tea aroma (e.g., linalool, geraniol, linalool oxide, and Trans-2- hexenal) during its fermentation process [10]. In addition, micro/nano-acoustic biosensors are frequently used to enhance the activity of specific biomolecules such as enzymes for increasing detection sensitivity [11]. These biosensors are based on a unique class of air-filled protein nanostructures called gas vesicles that vibrate in response to US waves. The use of US can easily image deep tissue with high spatiotemporal resolution. For instance, Jiang, et al. [12] used US for the bio-imaging of plant chemical composition by using quantum dots technology for in vitro cell imaging and the in vivo imaging of natural plants. Moreover, Lakshmanan, Jin, Nety, Sawyer, Lee-Gosselin, Malounda, Swift, Maresca and Shapiro [8] used acoustic biosensors for imaging the enzyme activity inside the mouse gastrointestinal tract. The principle of using acoustic-based biosensors is based on coupling the measurement nature (such as analyte adsorption) as a modulation in the physical properties of the acoustic wave (such as US frequency and velocity) that could be correlated with the analyte concentration [11]. Existing molecular biosensors, based on fluorescent emission, have limited utility due to the scattering of light and interference with their phytochemicals' fluorescents. The use of US can easily image deep tissue with high spatiotemporal resolution. Jiang, Jin and Gui [12] used a US-assisted solvothermal reaction for bio-imaging of plant zinc-ions by using quantum dots technology. The authors suggested that the viability of the technique could be used for in-vitro cell imaging and in vivo imaging of natural plants.

Furthermore, other emerging technologies have been developed for enhancing analytical measurement efficiency. For instance, Gouda, Chen, Li, Liu and He [1] fabricated an electrochemical method based on single plant cells for tracking the chemical composition and the antioxidant activity during the cultivation process. In addition, recent emerging and chemical-free technologies that are related to the in situ detection of the physicochemical changes in the biological media are one of this Special Issue's targets. As an example, Gouda, et al. [13] developed a method, based on Raman microspectroscopy and circular dichroism, for tracking the changes in the secondary protein structure of microalgae species and its impact on the physicochemical patterns. Moreover, several studies have documented the efficacy of these technologies for the replacement, enhancement, and improvement of various conventional analytical techniques in detecting animal and plant tissues [6,14–18].

Thus, the objective of this Special Issue is to demonstrate the potential of US and other recent physicochemical analytical technologies in providing a comprehensive chemical composition and bioactivity relationship of the different biological and organic chemicals. The topic collection includes, but is not limited to, molecular mechanisms of action of organic and inorganic molecules, especially if giving support to visualization approaches by acoustic-based sensors and biosensors, for example, identifying enzymes' biomarkers, as well as methodologies to investigate the chemical hazardous pollutants and heavy metals through sonochemistry and other related approaches. Further topic includes food, the environment, biomedicine, biotechnology, and the chemical composition of biosystems. In conclusion, this Special Issue could play an important role in maximizing the phytochemical functionality tracking and detection in the drug discovery and biotechnology fields through a very simple application via sonochemistry, electrochemistry, spectroscopy, and other related applications.

Funding: This work received no external funding.

Institutional Review Board Statement: Not applicable.

Informed Consent Statement: Not applicable.

Data Availability Statement: Not applicable.

Acknowledgments: The authors would like to thank MDPI for facilitating the launch of this Special Issue.

Conflicts of Interest: The authors declare no conflict of interest.

References

1. Gouda, M.; Chen, K.; Li, X.; Liu, Y.; He, Y. Detection of microalgae single-cell antioxidant and electrochemical potentials by gold microelectrode and raman micro-spectroscopy combined with chemometrics. *Sens. Actuators B Chem.* **2021**, *329*, 129229. [CrossRef]
2. Wang, Y.; Zhou, S.; Wang, M.; Liu, S.; Hu, Y.; He, C.; Li, P.; Wan, J.B. Uhplc/q-tofms-based metabolomics for the characterization of cold and hot properties of chinese materia medica. *J. Ethnopharmacol.* **2016**, *179*, 234–242. [CrossRef] [PubMed]
3. Dang, Z.; Liu, X.; Wang, X.; Li, M.; Jiang, Y.; Wang, X.; Yang, Z. Comparative effectiveness and safety of traditional chinese medicine supporting qi and enriching blood for cancer related anemia in patients not receiving chemoradiotherapy: A meta-analysis and systematic review. *Drug Des. Dev. Ther.* **2019**, *13*, 221–230. [CrossRef] [PubMed]
4. Gouda, M.; Sheng, L.; Aadil, R.M.; Liu, Y.; Ma, M.; Li, X.; He, Y.; Munekata, P.E.S.; Lorenzo, J.M. Interaction of bioactive mono-terpenes with egg yolk on ice cream physicochemical properties. *Foods* **2021**, *10*, 1686. [CrossRef] [PubMed]
5. Lv, J.M.; Gouda, M.; Zhu, Y.Y.; Ye, X.Q.; Chen, J.C. Ultrasound-assisted extraction optimization of proanthocyanidins from kiwi (actinidia chinensis) leaves and evaluation of its antioxidant activity. *Antioxidants* **2021**, *10*, 1317. [CrossRef] [PubMed]
6. Lv, J.-M.; Gouda, M.; El-Din Bekhit, A.; He, Y.-K.; Ye, X.-Q.; Chen, J.-C. Identification of novel bioactive proanthocyanidins with potent antioxidant and anti-proliferative activities from kiwifruit leaves. *Food Biosci.* **2022**, *46*, 101554. [CrossRef]
7. Gouda, M.; El-Din Bekhit, A.; Tang, Y.; Huang, Y.; Huang, L.; He, Y.; Li, X. Recent innovations of ultrasound green technology in herbal phytochemistry: A review. *Ultrason. Sonochemistry* **2021**, *73*, 105538. [CrossRef] [PubMed]
8. Lakshmanan, A.; Jin, Z.; Nety, S.P.; Sawyer, D.P.; Lee-Gosselin, A.; Malounda, D.; Swift, M.B.; Maresca, D.; Shapiro, M.G. Publisher correction: Acoustic biosensors for ultrasound imaging of enzyme activity. *Nat. Chem. Biol.* **2020**, *16*, 1035. [PubMed]
9. Bourdeau, R.W.; Lee-Gosselin, A.; Lakshmanan, A.; Farhadi, A.; Kumar, S.R.; Nety, S.P.; Shapiro, M.G. Acoustic reporter genes for noninvasive imaging of microorganisms in mammalian hosts. *Nature* **2018**, *553*, 86–90. [CrossRef]
10. Sharma, P.; Ghosh, A.; Tudu, B.; Sabhapondit, S.; Baruah, B.D.; Tamuly, P.; Bhattacharyya, N.; Bandyopadhyay, R. Monitoring the fermentation process of black tea using qcm sensor based electronic nose. *Sens. Actuators B Chem.* **2015**, *219*, 146–157. [CrossRef]
11. Fogel, R.; Limson, J.; Seshia, A.A. Acoustic biosensors. *Essays Biochem* **2016**, *60*, 101–110. [PubMed]

12. Jiang, X.; Jin, H.; Gui, R. Visual bio-detection and versatile bio-imaging of zinc-ion-coordinated black phosphorus quantum dots with improved stability and bright fluorescence. *Biosens. Bioelectron.* **2020**, *165*, 112390. [CrossRef] [PubMed]
13. Gouda, M.; Huang, Z.; Liu, Y.; He, Y.; Li, X. Physicochemical impact of bioactive terpenes on the microalgae biomass structural characteristics. *Bioresour. Technol.* **2021**, *334*, 125232. [CrossRef] [PubMed]
14. Chu, H.; Zhang, C.; Wang, M.; Gouda, M.; Wei, X.; He, Y.; Liu, Y. Hyperspectral imaging with shallow convolutional neural networks (scnn) predicts the early herbicide stress in wheat cultivars. *J. Hazard. Mater.* **2022**, *421*, 126706. [CrossRef] [PubMed]
15. Zhao, Y.; Zhang, J.; Gouda, M.; Zhang, C.; Lin, L.; Nie, P.; Ye, H.; Huang, W.; Ye, Y.; Zhou, C.; et al. Structure analysis and non-invasive detection of cadmium-phytochelatin2 complexes in plant by deep learning raman spectrum. *J. Hazard. Mater.* **2022**, *427*, 128152. [CrossRef] [PubMed]
16. Rehman, K.u.; Gouda, M.; Zaman, U.; Tahir, K.; Khan, S.U.; Saeed, S.; Khojah, E.; El-Beltagy, A.; Zaky, A.A.; Naeem, M.; et al. Optimization of platinum nanoparticles (ptnps) synthesis by acid phosphatase mediated eco-benign combined with photocatalytic and bioactivity assessments. *Nanomaterials* **2022**, *12*, 1079. [CrossRef] [PubMed]
17. Taha, M.F.; Abdalla, A.; ElMasry, G.; Gouda, M.; Zhou, L.; Zhao, N.; Liang, N.; Niu, Z.; Hassanein, A.; Al-Rejaie, S.; et al. Using deep convolutional neural network for image-based diagnosis of nutrient deficiencies in plants grown in aquaponics. *Chemosensors* **2022**, *10*, 45. [CrossRef]
18. Zong, W.; Gouda, M.; Cai, E.; Wang, R.; Xu, W.; Wu, Y.; Munekata, P.E.S.; Lorenzo, J.M. The antioxidant phytochemical schisandrin a promotes neural cell proliferation and differentiation after ischemic brain injury. *Molecules* **2021**, *26*, 7466. [CrossRef] [PubMed]

molecules

Article

Accumulation of Bisphenol A® by *Pleurotus* spp.—Flow Injection Analysis

Agata Krakowska [1,2,*], Małgorzata Suchanek [2], Robert Piech [2,*], Beata Paczosa-Bator [2], Tomasz Skalski [3] and Bożena Muszyńska [4]

1 Department of Inorganic Chemistry and Pharmaceutical Analytics, Faculty of Pharmacy, Jagiellonian University Medical College, 9 Medyczna Street, 30-688 Kraków, Poland
2 Department of Analytical Chemistry and Biochemistry, Faculty of Materials Science and Ceramics, AGH University of Krakow, Al. Mickiewicza 30, 30-059 Krakow, Poland; msuchanek@agh.edu.pl (M.S.); paczosa@agh.edu.pl (B.P.-B.)
3 Tunneling Group, Biotechnology Center, Silesian University of Technology, Bolesława Krzywoustego 8, 44-100 Gliwice, Poland; tomasz.skalski@polsl.pl
4 Department of Pharmaceutical Botany, Faculty of Pharmacy, Jagiellonian University Medical College, 9 Medyczna Street, 30-688 Kraków, Poland; muchon@poczta.fm
* Correspondence: agata.krakowska@uj.edu.pl (A.K.); rpiech@agh.edu.pl (R.P.); Tel.: +48-12-617-24-73 (A.K.)

Abstract: A specific feature of mushrooms (including those of the genus *Pleurotus*) is their natural ability to absorb and accumulate many chemical substances present in their immediate environment, which makes them an excellent natural sorption material. Hence, fruiting bodies of mushrooms have been recognized for years as excellent indicators of the environment, reflecting its current state. Nevertheless, mushrooms can accumulate both health-promoting substances, such as bioelements, and toxic substances, such as heavy metals and organic compounds, including bisphenol A® (BPA). This organic chemical compound in the phenol group, although it has been withdrawn in the EU since 2010, is widely present in the environment around us. In the present experiment, we aimed to determine the effect of adding BPA to liquid media for in vitro cultures of *Pleurotus* spp. The biomass increases were determined. Moreover, the degrees of adsorption and desorption of BPA from the obtained freeze-dried biomass in two different environments (neutral and acidic) were determined as a function of time. This is the first study to determine the bioavailability of adsorbed BPA in obtained biomass by extracting the mycelium into artificial digestive juices in a model digestive system. BPA was added to the liquid Oddoux medium in the following amounts: 0.01, 0.5, and 0.5 g/250 mL of medium. The amounts of adsorbed and desorbed BPA were determined by flow injection analysis (FIA) with amperometric detection. The addition of BPA to the substrate reduced the biomass growth in each of the discussed cases. BPA adsorption by the mycelium occurred at over 90% and depended on the morphology of the mushroom (structure, surface development, and pore size). BPA desorption depended on the pH of the environment and the desorption time. Mushrooms are an excellent natural remedial material, but BPA is extracted into artificial digestive juices; therefore, consuming mushrooms from industrialized areas may have health consequences for our bodies.

Keywords: accumulation; adsorption; desorption; bisphenol A®; *Pleurotus* spp.; extraction; artificial digestive juices; flow injection analysis

1. Introduction

The specific ability of higher mushrooms, including *Pleurotus* species, to accumulate has been the subject of numerous studies and scientific publications, especially in recent years [1–7]. The first reports on bioaccumulation concerning the sorption of some metals by mushrooms appeared nearly 100 years ago [8,9]. Initially, the focus was on assessing the toxicity of mushrooms resulting from the accumulation of heavy metals. A little later, the practical possibilities of using mushrooms as natural dietary ingredients enriched with selected substances began to be noticed [10–13]. Research on the specific properties of mushrooms gained importance after World War II, when attention began to be paid to their extraordinary features, which include their natural ability to effectively absorb and accumulate all chemical entities present in their immediate environment. Hence, fruiting bodies of "wild" mushrooms have been recognized for years as excellent environmental bioindicators [14–17] and used for mycoremediation. The uptake of substances by mushrooms is closely related to their availability in the substrate [18–20]. The presented work focused on the use of mycelial mushrooms as a natural material with specific sorption properties for bisphenol A®. This organic chemical compound in the phenol group was one of the most widely produced chemicals in the world [21], which is why it is widely distributed in the environment, even though it has been withdrawn in the European Union since 2010 [22]. It was used in the production of plastics, including polycarbonate, polyesters (present in baby feeding bottles and thermal papers for printers) [23–25], polysulfones, and epoxy resins, used for lining metal cans for canning and storing food, toys, drinking containers, sports equipment, medical equipment, and dental monomers, among others. Moreover, it was used in the food industry as an antioxidant and in cheaper cosmetics [23]. Studies have shown that despite its useful properties, this compound is toxic. It causes inflammation and numerous diseases of the genital organs (the ovaries, prostate, and endometrium [26]) and breasts. It has also been proven that BPA has an adverse effect on human reproduction by disturbing the functioning of the ovaries and their functioning during in vitro fertilization [27], deteriorating the properties of semen [28], and hindering the implantation of embryos [29]. Additionally, high levels of BPA in the body disturb the functioning of the thyroid gland [30,31]. Therefore, in the presented work, research was carried out on the accumulation in mycelial mushrooms of this dangerous compound for humans, which is still widely present in our environment. For this purpose, in vitro cultures of three species of mushrooms of the *Pleurotus* species (*P. djamor*, *P. pulmonarius*, and *P. ostreatus*) were established in media with the addition of bisphenol A®. There were several reasons for selecting these species for this research. These species belong to saprotrophic woody mushrooms, and unlike mycorrhizal mushrooms, they are much more active in the decomposition of chemical substances (including toxic substances) due to their high content of lignocellulolytic enzymes. Moreover, the specific structure of the hyphae of these mushrooms means that they have already been used, for example, to remove heavy metals, including Cd, Pb, or As [32]. Additionally, these mushrooms are obtained from crops and have a great taste and numerous health-promoting properties for humans (containing minerals, carbohydrates, phenolic and indole compounds [33], and acidic polysaccharides [34]). The influence of the addition of bisphenol A® in various amounts (0.05, 0.1, and 0.5 g of BPA/250 mL of medium) on its accumulation in the mycelium and biomass growth was examined. The efficiency of its adsorption and the degree of leaching of this compound depending on the time under the influence of two environmental factors (water and acid precipitation) were determined. Moreover, the influence of the mushroom's structural morphology (surface development, size, and number of pores) on the degree of BPA accumulation was determined (using SEM and BET analysis). BPA, which is a component of many plastics, is still present in the environment due to the fact that the decomposition of plastics is a long-term process. Therefore, alternative activities to support this process are being sought. One of the effective solutions is the use of biodegradation. Therefore, the conducted research aimed to demonstrate the possibility of using mycelium as a natural material for the accumulation of BPA. However, at the stage after the sorption of BPA by the mycelium, this material would require disposal because BPA is desorbed from it. Therefore, its consumption would pose a health risk to humans. This was confirmed by our research, in which the obtained biomasses of the tested mushroom species were extracted into artificial gastric and intestinal digestive juices in a model digestive system. This made it possible to determine the degree of risk (the amount of bisphenol A®) posed by the entry of this dangerous substance into the body when consumed in mushroom material. The analysis was carried out using the flow injection analysis (FIA) method. The obtained results broaden the knowledge of BPA accumulation by the mycelium of mushrooms and may contribute to the development of elimination and biodegradation methods for BPA.

2. Materials and Methods

2.1. The Scheme of the Experiment

This research was carried out according to the scheme presented in Figure 1.

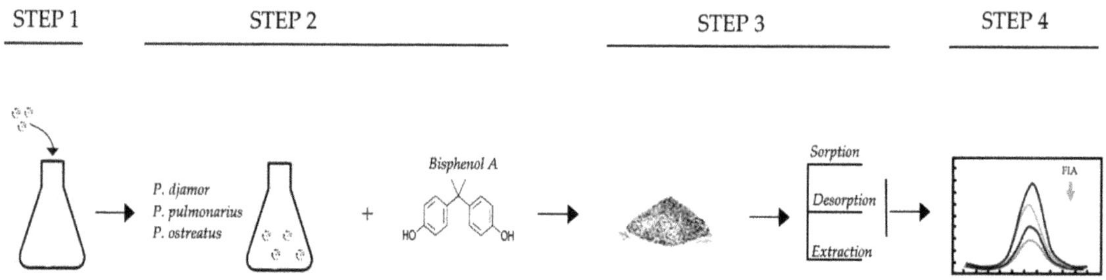

Figure 1. Scheme of the conducted experiment.

2.2. Mushroom Materials

2.2.1. Initial Mycelial Cultures (Step 1)

Mycelial cultures of three species of mushrooms of the *Pleurotus* genus were selected for the experiment: *Pleurotus djamor* (Rumph. ex Fr.), *Pleurotus ostreatus* (Jacq.), and *Pleurotus pulmonarius* (Fr.), which were obtained from an agar medium. The obtained biomasses from the in vitro cultures on solid media were transferred in an amount of 0.1 g of the inoculum into 500 mL Erlenmeyer flasks filled with 250 mL of a liquid medium (Oddoux medium (1957)). The in vitro cultures were performed while maintaining physicochemical conditions that were optimized in previous experiments [35]. Representative samples of the mycelia were deposited in the Department of Pharmaceutical Botany of the Jagiellonian University Medical College.

2.2.2. Experimental Mycelial Cultures (Step 2)

The experimental materials consisted of *Pleurotus* species biomasses obtained in the liquid Oddoux medium with the addition of bisphenol A® in an amount of 0.05, 0.1, or 0.5 g/250 mL of the medium (0.88 mM, 1.75 m, or 8.76 mM). The weighed amounts of bisphenol A® were dissolved in 1 mL of methanol using an ultrasonic bath (45 min) (Emag Emmi 20 HC, Mörfelden-Walldorf, Germany) and then quantitatively transferred to flasks.

In vitro shaken cultures were performed in 500 mL Erlenmeyer flasks with three independent replicates. After 14 days of mycelial growth, the mushroom material obtained from each tested species was separated from the medium by filtration (Pyrex Buchner funnel with a perforated plate, Merck, Darmstadt, Germany). The obtained biomass was freeze-dried (Freezone 4.5 freeze dryer, Labconco, Kansas, MO, USA; temperature: −40 °C).

2.3. Mushroom Material Lyophilization Analysis (Step 3)

2.3.1. Sorption and Desorption

The obtained freeze-dried biomasses of the three *Pleurotus* species (step 1) from the experiment were subjected to a sorption and desorption study for bisphenol A® in order to determine the degree of its connection with the matrix. The degree of sorption was determined by analyzing the difference between the content of bisphenol A® in the initial medium for the shaken mycelial cultures to which it was added in an amount of 0.05, 0.1, or 0.5 g/250 mL of the medium (0.88 mM, 1.75 mM, or 8.76 mM) and its amount remaining in the medium after 14 days of culture. In turn, in order to determine the degree of desorption, amounts of 0.5 g of the obtained biomasses were transferred to 50 mL of a solution. Desorption tests were carried out in two environments: an aqueous environment and that corresponding to acid rain conditions (using a 10% nitric acid solution, for which 65% nitric acid was diluted in quadruple-distilled water). The analysis was performed in three independent repetitions each time.

2.3.2. Extraction of Bisphenol A® into Artificial Digestive Juices (Step 3)

Preparation of Artificial Digestive Juices

In order to carry out the planned experiment, solutions of artificial digestive juices (saliva, gastric juice, and intestinal juice) were prepared. The individual ingredients of the digestive juices are presented in Table 1. All ingredients were weighed with an accuracy of 0.1 mg with an analytical scale. The components of the individual digestive juices weighed in this way were transferred to flasks with a volume of 1000 mL, which were filled to the mark with quadruple-distilled water.

Table 1. Compositions of artificial digestive juices [36–38].

	Chemical Compounds	Weight (g)/L	Direction of extraction
Saliva (pH = 6.7)	$CaCl_2$	0.20	↓
	$C_6H_8O_7$	0.03	
	$KHCO_3$	1.50	
	KH_2PO_4	0.35	
	$MgCl_2$	0.01	
	Na_2HPO_4	0.35	
Gastric juice (pH = 2)	HCl	0.10	
	NaCl	2.00	
	Pepsin	3.20	
Intestinal juice (pH = 8)	Bile salt	0.15	
	$NaHCO_3$	8.50	
	Pancreatic extract	0.02	

Extraction Process

The obtained biomasses (step 2) were homogenized with an analytical mill (EGK, Rommelsbacher, Plauen, Germany). The materials thus obtained were weighed to an amount of 0.5 g and extracted in the artificial digestive juices. The process was carried out in three independent repetitions. For this purpose, the weighed amounts of the biomasses were placed in flat-bottomed flasks. Then, the mushroom materials were moistened with 3 mL of artificial saliva. An amount of 20 mL of artificial gastric juice was added to the materials. The mixtures prepared in this way were incubated for 60 min in a Gastroel-2014 apparatus (Krakow, Poland) [39]. This device allows extraction tests to be carried out in physicochemical conditions similar to those prevailing in the human body (37 °C). After a period of 1 h, the contents of the flat-bottomed flasks were separated from the filtrates by filtration using membrane paper (Ø 0.22 µm, Millex, Millipore Corporation, Burlingtone, MA, USA). In the next step, 20 mL of intestinal juice was added to the separated biomasses, and the digestion process was continued for 150 min so as to maintain the natural sequence of the digestion process and the time corresponding to the physiological period of stay of food content in the human digestive system. After this period, the filtration process was repeated. The bisphenol A® content was determined in the obtained filtrates using the FIA method.

2.4. Reagents

In the presented experiment, the bisphenol A® standard was purchased from Merck (Darmstadt, Germany, cat. no.: 80057). Cultures: $ZnCl_2$ (cat. no.: 229997) and $MgCl_2$ (cat. no.: 63138) were purchased from Sigma-Aldrich (Darmstadt, Germany). The ingredients of the artificial digestive juices: NaCl and $NaHCO_3$ were obtained from PPH Golpharm

(Kraków, Poland); pepsin and bile salts were obtained from BTL (Łódź, Poland); $CaCl_2$ was obtained from Pharma Zentrale GmbH (Herdecke, Germany); pancreatic extract, HCl, KCl, concentrated HNO_3, Suprapur®, and KNO_3 were purchased from Merck (Darmstad, Germany); and $C_6H_8O_7$, $KHCO_3$, Na_2HPO_4, K_2HPO_4, KH_2PO_4, and NaOH were acquired from the Polish Society of Chemistry (Gliwice, Poland). The HPLC-grade methanol used to dissolve bisphenol A® was purchased from Sigma-Aldrich (St. Louis, MO, USA). The quadruple-distilled water used with a conductivity of less than 1 μS/cm was obtained using an S2-97A2 distillation apparatus (ChemLand, Stargard Szczecin, Poland).

2.5. Analytical Tools Applied

In the presented experimental work on the mycoremediation of bisphenol A® by the mycelia of the Pleurotus species, the degree of adsorption, desorption, and extraction of bisphenol A® into plastic digestive juices was determined using the measurement technique of FIA. The visualization of the characteristics and surface development of the materials (the mycelia of the Pleurotus species) was performed using scanning electron microscopy (SEM) and the Brunauer–Emmett–Teller isotherm (BET).

2.5.1. Surface Characteristics

The microstructures and surface morphologies of the *P. djamor*, *P. pulmonarius*, and *P. ostreatus* mycelia (the homogenized biomasses from the in vitro cultures) were observed under an SEM microscope (FEI Nova NanoSem 200, Thermo Fisher Scientific, Hillsboro, OR, USA) with an accelerating voltage of 18 kV and magnification of 3.000×. The surface areas and total pore volumes of the three mycelia (*P. djamor*, *P. pulmonarius*, and *P. ostreatus*) were measured by N2 adsorption using the Brunauer–Emmett–Teller (BET) and Langmuir methods (ASAP 2010, Micromeritics, Norcross, GA, USA).

2.5.2. Bisphenol A® Content Analysis (FIA Detection) (Step 4)

The flow system was composed of a 0.05 L K_2HPO_4 (0.01 M) solution reservoir and an 800 Dosino pump connected to a 900 Touch Control panel (Metrohm, Herisau, Switzerland), as well as a sample injection valve (Rheodyne Model 7010), a flow wall-jet detector, and a valve with a sample loop with a volume of 100 μLA. Screen-printed carbon electrodes (SPCEs) with built-in reference electrodes (Metrohm, Herisau, Switzerland) were used as well. Amperometric determination of BPA under flow injection conditions was performed. The characteristic parameters (working potential and flow rate) of the FIA were investigated. The optimal potential was chosen as 700 mV, and the calibrations were registered with an optimal flow rate of 0.01 M K_2HPO_4 equal to 1.0 mL/min. The signals for BPA determination under flow conditions were registered in a BPA concentration range of 10 to 30 μM. The BPA concentration in the medium was determined under flow analysis conditions. The measurements were carried out using the calibration method. The calibration curves were registered in 0.01 M K_2HPO_4 containing 50 μL of the medium. The calibration curve for BPA determination in the range of 10 to 30 μM by FIA is presented in Figure 2. The obtained slope of the regression line for BPA was 0.0142 ± 0.003 μA/μM (intercept: 0.0533 ± 0.0051 μM; r = 0.9997).

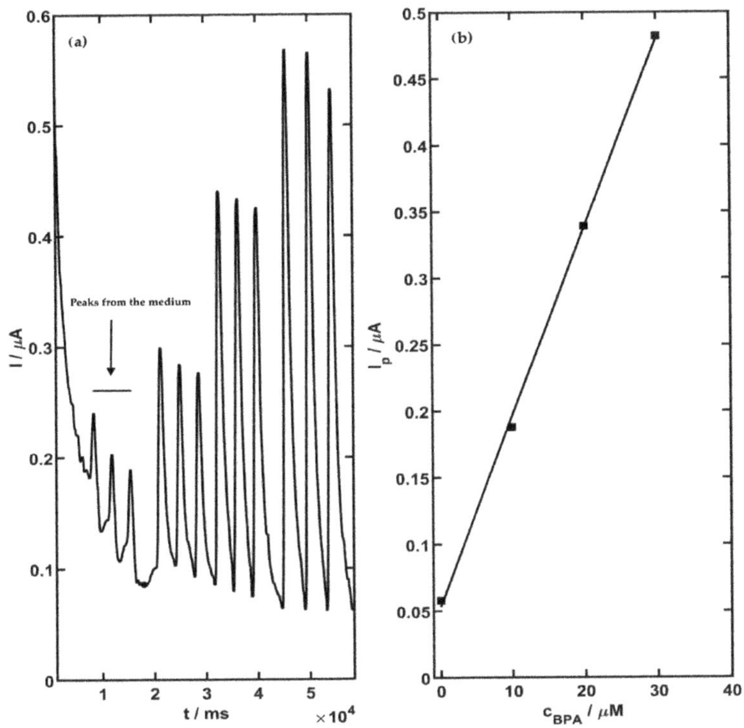

Figure 2. 2,2-bis (p-hydroksyfenylo)propan calibration chart (**a**) and curve (**b**) for concentration range from 10 to 30 μM using screen-printed carbon electrodes with amperometric parameters of measurements under flow injection conditions.

3. Results and Discussion

In the presented research work, the degree of adsorption of bisphenol A® dosed into the substrate (culture medium) was analyzed for three amounts (0.05, 0.1, and 0.5 g/250 mL of medium) depending on the species of the *Pleurotus* mycelia (*P. djamor*, *P. ostreatus*, and *P. pulmonarius*). In turn, the obtained biomasses were subjected to desorption analysis, which was carried out in two environments (neutral and acidic) at three time points (1 h, 2 h, and 6 h), and extraction in artificial digestive juices. The obtained natural materials (biomasses from in vitro cultivation) were characterized. The results are presented in the following subsections.

3.1. Material Characteristics

The SEM images of the *P. djamor*, *P. pulmonarius*, and *P. ostreatus* mycelia are shown in Figure 3. As can be seen, the *P. ostreatus* mycelium has an obvious microporous structure, which can provide a high specific surface area. The least porous structure is observed for the *P. djamor* mycelium, which can provide a low specific surface area.

By means of the BET method, it was found that the specific surface areas of the examined mycelia were 1.47 m^2/g for *P. djamor*, 1.92 m^2/g for *P. pulmonarius*, and 4.21 m^2/g for *P. ostreatus*. The surface areas obtained by the Langmuir equation were 2.19 m^2/g, 2.82 m^2/g, and 6.58 m^2/g for the *P. djamor*, *P. pulmonarius*, and *P. ostreatus* mycelia, respectively. The total micropore volumes of *P. djamor*, *P. pulmonarius*, and *P. ostreatus* were 0.25 mm^3/g, 0.197 mm^3/g, and 1.094 mm^3/g, respectively. The average pore diameters of the mycelia were 13.6 nm, 12.0 nm, and 12.1 nm for *P. djamor*, *P. pulmonarius*, and *P. ostreatus*, respectively.

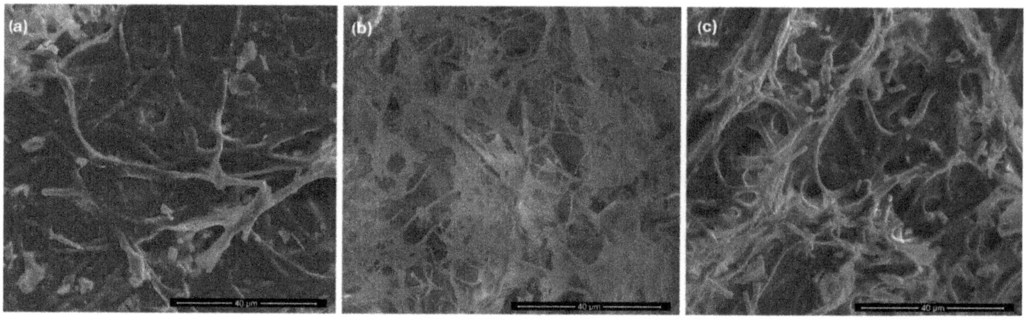

Figure 3. SEM images of *P. djamor* (**a**), *P. pulmonarius* (**b**), and *P. ostreatus* (**c**) mycelia.

3.2. Biomass Growth Analysis

The analysis of biomass growth was performed on freeze-dried and homogenized materials. For this purpose, the biomass obtained from the three independent cultures for each species and the addition of bisphenol A® was weighed with an analytical scale with an accuracy of four decimal places. The results are presented in Figure 4.

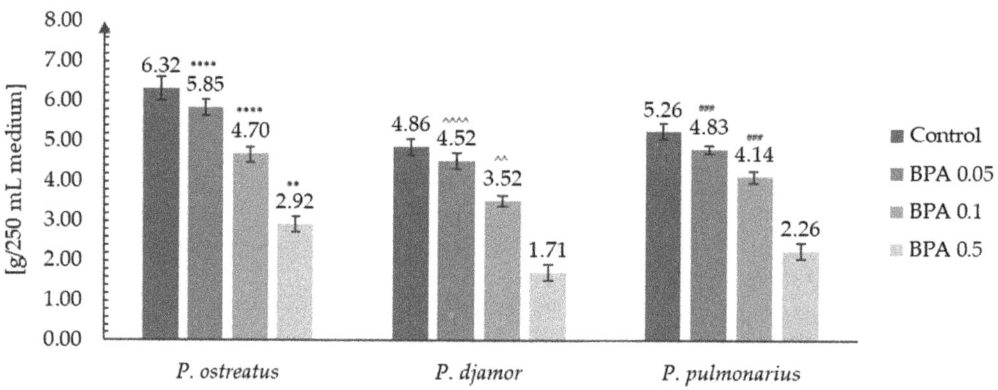

Figure 4. Biomasses obtained after in vitro cultures of *Pleurotus* species (*P. djamor, P. ostreatus,* and *P. pulmonarius*) in the liquid Oddoux medium (g/250 mL of medium) depending on the amount of bisphenol A® added (0.05, 0.1, or 0.5 g/250 mL of medium) (one-way ANOVA with post hoc Tukey's test: ** $p < 0.01$ and **** $p < 0.0001$ vs. *P. djamor* control; ^^ $p < 0.01$ and ^^^^ $p < 0.0001$ vs. *p. pulmonarius* control; ### $p < 0.001$ vs. *P. ostreatus* control).

The analysis showed that the addition of bisphenol A® to the medium, even in a small amount, reduced the growth of the mycelial biomass (Figure 4). The highest increase in biomass for the smallest addition of bisphenol A® (0.05 g/250 mL of medium) was obtained in the case of the in vitro culture of the *P. ostreatus* species (5.85 g of dry biomass/250 mL of medium). In turn, the lowest biomass, in this case, was observed in the in vitro cultures of the *P. djamor* species. The lowest biomass increase was recorded for the addition of 0.5 g of BPA to the medium. In this case, an over 50% decrease in biomass was observed for all *Pleurotus* species discussed in this work (Figure 4). Despite the observed decreases in biomass growth for the in vitro cultures of the *Pleurotus* species obtained in media with the addition of BPA, their values were still higher than, for example, in the case of the *P. pulmonarius* species obtained by Włodarczyk et al. in 2021 conducted in a medium enriched with $MgSO_4 \cdot 7H_2O$ salts (6.58 g/L of medium) (control: 2.89 g/L of medium) [35].

Similarly, in another experiment (Confortin et al., 2008) in which the production of mycelial biomass was optimized, a maximum of 5.49 g of mycelium per liter of substrate was obtained in the case of *Pleurotus* sajor-caju [40]. In turn, in other experiments in which the researchers added different concentrations of glucose to the media (Rosado et al., 2003), they achieved a higher biomass growth efficiency, which, in the extreme case, amounted to as much as 22.8 g of dry mycelium per 1 L of the medium [41,42].

3.3. Sorption Analysis

Flow injection analysis (FIA) with amperometric detection was used for the BPA determination in the medium before and after sorption. The operating parameters are given in Section 2.5.2. The FIA signal was recorded each time in three independent repetitions with the addition of the medium in the range of 5 to 50 µL. The obtained FIA curves for the medium before and after sorption by *P. djamor*, *P. pulmonarius*, and *P. ostreatus* are presented in Figure 5. The BPA concentration in the medium was calculated from the regression line using the following equation:

$$I\_p = 0.0143 \cdot c\_BPA + 0.0533$$

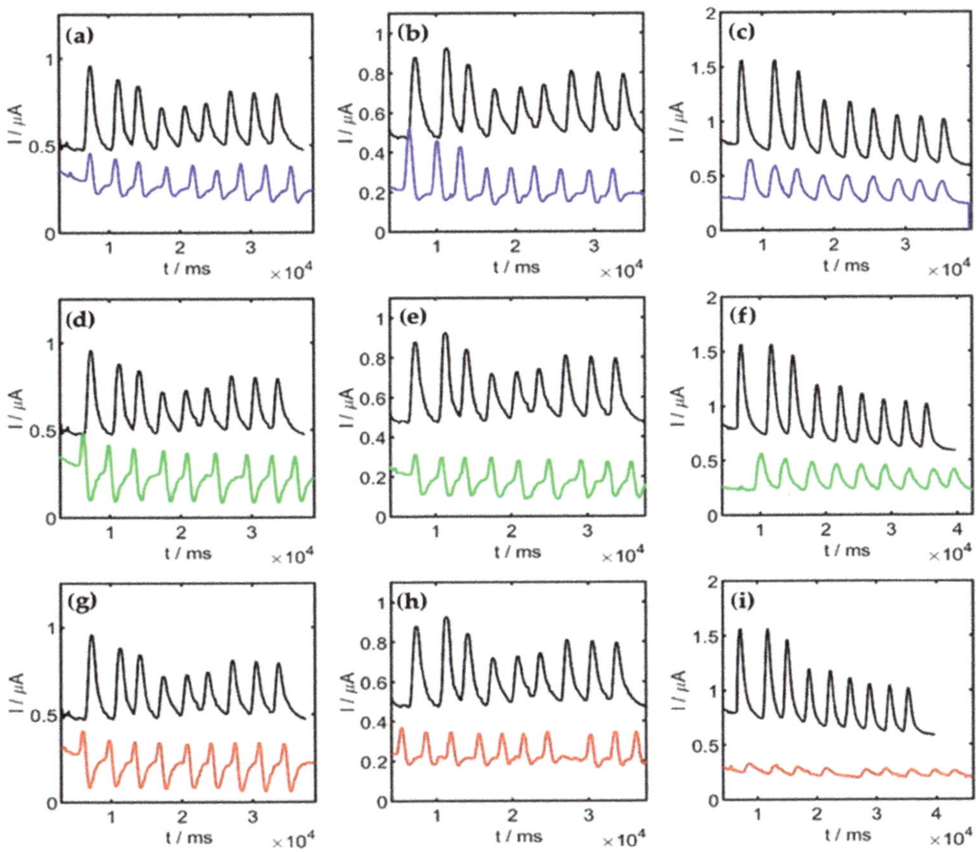

Figure 5. BPA signals registered in medium before (black) and after sorption of 0.05 g (**a**,**d**,**g**), 0.1 g (**b**,**e**,**h**), and 0.5 g (**c**,**f**,**i**) of BPA by *P. djamor* (blue), *P. pulmonarius* (green), and *P. ostreatus* (red).

Based on the sorption analysis, it was shown that mushrooms of the *Pleurotus* species effectively sorbed bisphenol A® from the substrate in each of the discussed cases (Table 2).

Table 2. Sorption efficiencies (shortcuts: *P. djamor*—Pdj, *P. pulmonarius*—Pp, and *P. ostreatus*—Po).

	Sorption of 0.05 g of BPA/250 mL of medium [%]
Pdj	92.8 ± 6.4
Pp	95.9 ± 3.1
Po	**99.1** ± 0.8
	Sorption of 0.1 g of BPA/250 mL of medium [%]
Pdj	91.3 ± 1.5
Pp	93.3 ± 1.2
Po	**99.0** ± 0.6
	Sorption of 0.5 g of BPA/250 mL of medium [%]
Pdj	65.3 ± 7.3
Pp	66.9 ± 4.6
Po	**97.5** ± 1.3

The analysis of the sorption efficiency showed that its degree decreased with the increase in the amount of BPA added to the medium. In the case of the addition of 0.05 g of BPA/250 mL of medium, the sorption efficiency for all species was over 90%. In turn, in the case of the addition of the highest concentration—0.5 g of BPA/250 mL of medium—a large disproportion was observed (Table 2). The species distinguished by the highest sorption efficiency was *P. ostreatus* (over 90% sorption). In turn, the lowest sorption efficiency was demonstrated by the *P. djamor* mycelium (65.3% sorption). This indicates that the sorption efficiency is related to the morphology of the mycelium, including its structure, surface development, size, and the shape of its pores. The species with the highest surface area and pore size and number had the best sorption, i.e., *P. ostreatus* (Section 3.1).

The tests carried out on the accumulation of BPA by mycelium show that this material is an excellent sorbent and can be an alternative to existing solutions. So far, BPA degradation has mainly involved recycling plastics, in which BPA is returned to the environmental cycle. Hence, the use of biodegradation is a more effective way of degrading plastics due to the environmentally friendly mechanism that reduces their pollution. Moreover, this process can be carried out under controlled in vitro conditions, as shown in this work. Mushrooms play a key role in the biodegradation of plastics by secreting degrading enzymes, i.e., cutinase, lipase, proteases, and lignocellulolytic enzymes. The action of these enzymes is based on the effective hydrolysis or oxidation of polymers, resulting in the formation of functional groups that improve their hydrophilic properties and, as a result, cause their decomposition. So far, several species of mushrooms are known whose degradative effect has been confirmed, e.g., *Aspergillus* species and *Cladosporium* species. These also include saprotrophic mushrooms, e.g., *Agaricus bisporus* and *Pleurotus* spp., in particular, *Pleurotus ostreatus*, which was the subject of the above work [43].

3.4. Desorption Analysis

In order to determine whether the bisphenol A® absorbed from the substrate during mycelial growth undergoes the reverse process, desorption, the obtained freeze-dried biomass was weighed to an amount of 0.5 g and, in three independent repetitions, introduced into 50 mL of a solution of quadruply distilled water (pH = 7) and an acidic solution, corresponding to acid precipitation in the natural environment (pH = 2.6). Desorption was carried out at three time points, 1, 3, and 6 h, after which samples were taken and analyzed using the FIA method. The obtained results are presented in Figure 6.

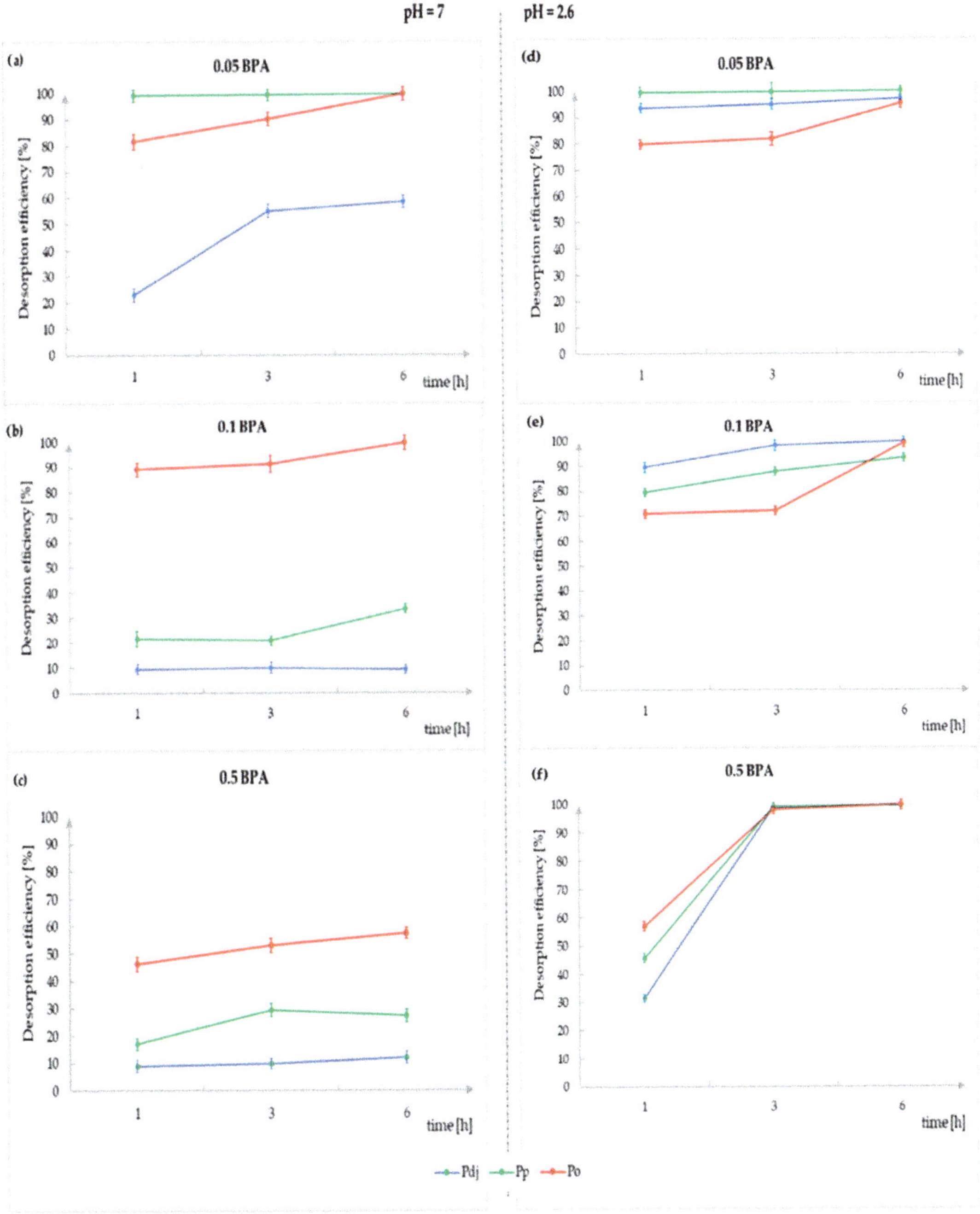

Figure 6. A graph presenting the desorption of BPA in a neutral environment with pH = 7 and acidic environment with pH = 2.6 for *Pleurotus* species mycelia (*P. djamor* (blue line), *P. pulmonarius* (green line), and *P. ostreatus* (red line)) depending on the desorption time (1, 3, or 6 h) and BPA additive to the medium (**a**,**d**—00.5, **b**,**e**—0.1, or **c**,**f**—0.5 g of BPA/250 mL of medium).

3.4.1. Desorption in Water

The analysis of BPA desorption in an aqueous environment (pH = 7) showed that for the mycelia obtained in a medium with the addition of 0.05 g of BPA/250 mL of medium, *P. djamor* was desorbed to the lowest degree (58.6%), and *P. pulmonarius* and *P. ostreatus* were desorbed to the highest and comparable degree (over 90%). Moreover, in the case of the first species mentioned, the degree of desorption increased with time. In turn, for the *P. pulmonarius* and *P. ostreatus* species, the amount of desorbed BPA stabilized after 3 h (Figure 6). However, in the case of the analysis of BPA desorption from the mycelia obtained in a medium with the addition of 0.1 g of BPA/250 mL of medium, a decrease in the desorption efficiency was observed for the *P. djamor* species (desorption after 6 h: 9.4%) and the *P. pulmonarius* species (desorption after 6 h: 33.3%). Only in the case of *P. ostreatus*, the decrease in desorption was small and amounted to approximately 1%. A completely different effect was observed in the case of the *P. ostreatus* mycelium obtained in a medium with the addition of 0.5 g of BPA/250 mL of medium. In this case, the degree of desorption dropped by over 40%, and after 6 h, it was 57.3%. Additionally, it was observed that the desorption carried out in an aqueous environment stabilized after 3 h, and the further desorption did not significantly increase its degree.

3.4.2. Desorption in Acidic Media

The study of BPA desorption in an acidic environment (pH = 2.6) showed that the pH significantly affected the desorption efficiency. In this case, desorption occurred to a higher degree for all of the discussed *Pleurotus* species. The degree of desorption increased to over 80% for the *P. djamor and P. pulmonarius* species in the case of the analysis of the desorption from the biomass obtained in a culture with the addition of BPA at a level of 0.1 g/250 mL of medium and up to over 90% in the case of adding 0.5 g/250 mL of medium. In turn, for the *P. ostreatus* species, the desorption from the biomass obtained for all three BPA additives was at a similar level and amounted to over 90%. Also, in the case of desorption into an acidic environment, as well as into a neutral-pH environment, the desorption time depended on the degree of desorption. Also, in this case, the degree of desorption stabilized after 3 h, and further incubation in the solution did not significantly increase its effectiveness (Figure 7b).

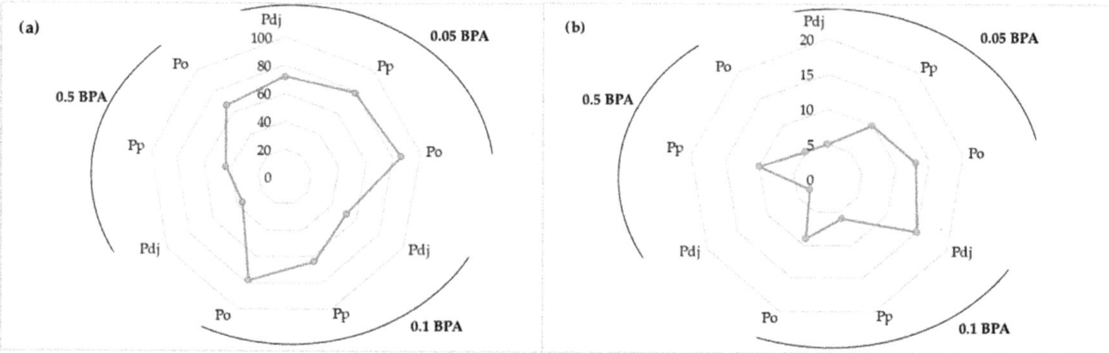

Figure 7. Graph showing the desorption of BPA into artificial digestive juices—(**a**) gastric juice and (**b**) intestinal juice (Pdj—*P. djamor*, Pp—*P. pulmonarius*, and Po—*P. ostreatus*)—from the biomass obtained from breeding with the addition of 0.05, 0.1, and 0.5 g/250 mL of BPA to the medium.

3.5. Extraction into Artificial Digestive Juices

The efficiency of bisphenol A® extraction from the obtained biomasses from the in vitro cultures of the *Pleurotus* species depending on the extraction site (gastric juice or intestinal juice) is presented in Figure 7a,b.

Based on the extraction into artificial digestive juices, it was found that the degree of extraction depended on the place of extraction (gastric juice or intestinal juice) as well as on the species of mushroom (*P. djamor*, *P. pulmonarius*, or *P. ostreatus*). The highest extraction efficiency was observed in artificial gastric juice, at over 80%). This was a similar trend to that in the case of desorption in an acidic medium. In that case, the pH in the gastric juice was also acidic and was approximately two. Additionally, this reaction affected the destruction of the mycelial cell walls, which resulted in a larger amount of BPA released from the inside of the mycelium. The highest extraction efficiency was observed in the case of the *P. ostreatus* species, and it was 83% in the gastric juice solution (Figure 7a) and 11% in the intestinal juice (Figure 7b). However, extraction by the *P. djamor* and *P. pulmonarius* species occurred to a similar extent and was at a level of 45%. The extraction in the intestinal juice, the pH of which was slightly alkaline (pH = 8), ranged from 3 to 15%. The low level of extraction in this place was also influenced by the fact that most of the BPA from the mycelia was extracted earlier in the gastric juice. Although BPA is an organic substance, similar to the analysis of metals in mushrooms, its extraction into gastric juice is more effective, which is related to the pH that occurs in the gastric juice, which is strongly acidic [44].

4. Conclusions

The conducted research confirms that mycelial mushrooms accumulate chemical entities present in their environment (e.g., toxic bisphenol A®). Hence, they constitute a natural material with adsorption properties. Even though BPA has been withdrawn from use, it is still present in the environment around us. The efficiency of BPA adsorption is closely related to the morphology of the mushrooms, i.e., their surface development and the number and size of pores. In turn, the degree of desorption depends on the pH of the solution and is much greater in the case of solutions with a pH below seven, as demonstrated in this work. Moreover, it was shown that this compound was extracted from mushroom biomass into artificial digestive juices to a degree of over 80%, so consuming mushrooms from industrialized areas poses a potential threat to human health and life. Studies have shown that mycelium is a good material for the accumulation of BPA, but whether its mechanism is physisorption or chemisorption has not been specified. In the next stage, research work will be aimed at explaining this, which will enable us, in the future, to propose an effective method of eliminating BPA from the material and, consequently, reducing it in the natural environment.

Author Contributions: Conceptualization, A.K.; methodology, A.K.; validation, A.K. and R.P.; formal analysis, A.K. and M.S.; investigation, A.K.; resources, A.K., B.M., R.P. and B.P.-B.; writing—original draft preparation, A.K.; writing—review and editing, A.K. and T.S.; visualization, A.K.; supervision, A.K.; project administration, A.K.; funding acquisition, A.K.; B.M., preparation of in vitro mycelial cultures. All authors have read and agreed to the published version of the manuscript.

Funding: This research was funded by Jagiellonian University (grant number N42/DBS/000012 and N42/DBS/000151).

Institutional Review Board Statement: Not applicable.

Informed Consent Statement: Not applicable.

Data Availability Statement: Data are contained within the article.

Conflicts of Interest: The authors declare no conflicts of interest.

References

1. Boamponsem, G.A.; Obeng, A.K.; Osei-Kwateng, M.; Badu, A.O. Accumulation of heavy metals by Pleurotus ostreatus from soils of metal scrap sites. *Int. J. Cur. Res. Rev.* **2013**, *5*, 1–9.
2. Ronda, O.; Grządka, E.; Ostolska, I.; Orzeł, J.; Cieślik, B.M. Accumulation of radioisotopes and heavy metals in selected species of mushrooms. *Food Chem.* **2022**, *367*, 130670. [CrossRef] [PubMed]
3. Brzostowski, A.; Falandysz, J.; Jarzyńska, G.; Zhang, D. Bioconcentration potential of metallic elements by Poison Pax (*Paxillus involutus*) mushroom. *J. Environ. Health* **2011**, *46*, 378–393. [CrossRef] [PubMed]

4. Falandysz, J.; Kojta, A.K.; Jarzyńska, G.; Drewnowska, A.; Dryżałowska, A.; Wydmańska, D.; Szefer, P. Mercury in Bay Bolete *Xerocomus badius*: Bioconcentration by fungus and assessment of element intake by humans eating fruiting bodies. *Food Addit. Contam.* **2012**, *29*, 951–961. [CrossRef]
5. Mleczek, M.; Siwulski, M.; Stuper-Szablewska, K.; Rissmann, I.; Sobieralski, K.; Goliński, P. Accumulation of elements by edible mushroom species. Problem of trace element toxicity in mushrooms. *J. Environ. Sci. Health* **2013**, *28*, 69–81. [CrossRef] [PubMed]
6. Javaid, A.; Rukhsana Bajwa, R.; Shafique, U.; Anwar, J. Removal of heavy metals by adsorption on Pleurotus ostreatus. *Biomass Bioenergy* **2011**, *35*, 1675–1682. [CrossRef]
7. Byrne, A.R.; Ravnik, V.; Kosta, L. Trace element concentrations in higher fungi. *Sci. Total Environ.* **1976**, *6*, 65–78. [CrossRef] [PubMed]
8. Stijve, T.; Besson, R. Mercury, cadmium, lead, and selenium content of mushroom species belonging to the genus Agaricus. *Chemosphere* **1976**, *5*, 151–158. [CrossRef]
9. Muszyńska, B.; Kała, K.; Włodarczyk, A.; Krakowska, A.; Ostachowicz, B.; Gdula-Argasińska, J.; Suchocki, P. *Lentinula edodes* as a source of bioelements released into artificial digestive juices and potential anti-inflammatory material. *Biol. Trace Elem. Res.* **2020**, *194*, 603–613. [CrossRef]
10. Zięba, P.; Kała, K.; Włodarczyk, A.; Szewczyk, A.; Kunicki, E.; Sękara, A.; Muszyńska, B. Selenium and zinc biofortification of *Pleurotus eryngii* mycelium and fruiting bodies as a tool for controlling their biological activity. *Molecules* **2020**, *25*, 889. [CrossRef]
11. Krakowska, A.; Zięba, P.; Włodarczyk, A.; Kała, K.; Sułkowska-Ziaja, K.; Bernaś, E.; Sękara, A.; Ostachowicz, B.; Muszyńska, B. Selected edible medicinal mushrooms from *Pleurotus* genus as an answer for human civilization diseases. *Food Chem.* **2020**, *327*, 127084. [CrossRef]
12. Reczyński, W.; Muszyńska, B.; Opoka, W.; Smalec, A.; Sułkowska-Ziaja, K.; Malec, M. Comparative study of metals accumulation in cultured in vitro mycelium and naturally grown fruiting bodies of *Boletus badius* and *Contharellus cibarius*. *Biol. Trace Elem. Res.* **2013**, *155*, 355–362. [CrossRef]
13. Krakowska, A.; Reczyński, W.; Muszyńska, B. Optimization of the liquid culture medium composition to obtain the mycelium of *Agaricus bisporus* rich in essential minerals. *Biol. Trace Elem. Res.* **2016**, *173*, 231–240. [CrossRef]
14. Yamanaka, K. Mushroom cultivation in Japan. *Mushroom Prod. Bull.* **2011**, *10*, 455–459.
15. Gapiński, M.; Woźniak, W.; Ziobra, M. Boczniak Technologia Uprawy i Przetwarzania. *Państwowe Wydaw. Rol. I Leśne* **2001**, *23*, 12–15.
16. Wondratschek, I.; Röder, U. Monitoring of heavy metals in soils by higher fungi. In *Plants as Biomonitors, Indicators for Heavy Metals in the Terrestrial Environment*; Markert, B., Ed.; Wiley-Blackwell: Hoboken, NJ, USA, 1993; pp. 345–363.
17. Reider, S.R.; Brunner, I.; Horvat, M.; Jacobs, A.; Frey, B. Accumulation of mercury and methylmercury by mushrooms and earthworms from forest soils. *Environ. Pol.* **2011**, *159*, 2861–2869. [CrossRef]
18. Svoboda, L.; Chrastný, V. Levels of eight trace elements in edible mushrooms from a rural area. *Food Additiv. Contam.* **2008**, *25*, 51–58. [CrossRef]
19. Campos, J.A.; Tejera, N.A.; Sánchéz, C.J. Substrate role in the accumulation of heavy metals in sporocarps of wild fungi. *Biometals* **2009**, *22*, 835–841. [CrossRef] [PubMed]
20. Svoboda, L.; Zimmermannová, K.; Kalač, P. Concentrations of mercury, cadmium, lead and copper in fruiting bodies of edible mushrooms in an emission area of a copper smelter and a mercury smelter. *Sci. Total Environ.* **2000**, *246*, 61–67. [CrossRef] [PubMed]
21. Burridge, E. Bisphenol A: Product profile. *Eur Chem. News.* **2003**, *17*, 14.
22. Konieczna, A.; Rutkowska, A.; Rachoń, D. Health risk of exposure to bisphenol A (BPA). *Rocz. Państwowego Zakładu Hig.* **2015**, *66*, 5–11.
23. Wetherill, Y.; Akingbemi, J.; Kanno, I.; Mclachlan, J.; Nadal, A.; Sonneschein, C.; Belcher, S. In vitro molecular mechanisms of bisphenol A action. *Reprod. Toxicol.* **2007**, *24*, 178. [CrossRef] [PubMed]
24. Rochester, J.R. Bisphenol A and Human Health: A review of the literature. *Reprod. Toxicol.* **2013**, *42*, 132–155. [CrossRef] [PubMed]
25. Geens, T. A review of dietary and non-dietary exposure to bisphenol A. *Food Chem. Toxicol.* **2012**, *50*, 3725–3740. [CrossRef] [PubMed]
26. Lang, I.A.; Galloway, T.S.; Scarlett, A.; Henley, W.E.; Depledge, M.; Wallace, R.B.; Melzer, D. Association of Urinary Bisphenol A Concentration With Medical Disorders and Laboratory Abnormalities in Adults. *JAMA* **2008**, *300*, 1303–1310. [CrossRef] [PubMed]
27. Ehrlich, S.; Williams, P.; Missmer, A.; Flaws, A.; Ye, X.; Calafat, A.; Hauser, R. Urinary bisphenol A concentrations and early reproductive health outcomes among women undergoing IVF. *Hum. Reprod.* **2012**, *27*, 3583. [CrossRef] [PubMed]
28. Bloom, M.; Vom Saal, F.; Kim, D.; Taylor, J.; Lamb, J.; Fujimoto, V. Serum unconjugated bisphenol A concentrations in men may influence embryo quality indicators during in vitro fertilization. *Environ. Toxicol. Pharmacol.* **2011**, *32*, 319. [CrossRef] [PubMed]
29. Ehrlich, S.; Williams, P.; Missmer, A.; Flaws, A.; Berry, K.; Calafat, A.; Hauser, R. Urinary bisphenol A concentrations and implantation failure among women undergoing in vitro fertilization. *Environ. Health. Perspect.* **2012**, *120*, 978. [CrossRef] [PubMed]
30. Meeker, J.; Calafat, A.; Hauser, R. Bisphenol A and thyroid hormones. *Environ. Sci. Technol.* **2010**, *44*, 1458. [CrossRef]
31. Wang, F.; Hua, J.; Chen, M.; Xia, Y.; Zhang, Q.; Zhao, R. Occupational exposure to bisphenol A (BPA) in a plastic injection molding factory in Malaysia. *Occupenviron. Med.* **2012**, *69*, 679.

32. Zhang, Y.; Chen, X.; Xie, L. *Pleurotus pulmonarius* Strain: Arsenic(III)/Cadmium(II) Accumulation, Tolerance, and Simulation Application in Environmental Remediation. *Int. J. Environ. Res. Public Health* **2023**, *20*, 5056. [CrossRef] [PubMed]
33. Matkovits, A.; Fodor, M.; Jókai, Z. Analysis of Polyphenol Patterns of *Pleurotus ostreatus* Cultivars by UHPLC-ESI-MS/MS; Application of FT-NIR and Chemometric Methods, Classification Options. *Chemosensors* **2024**, *12*, 19. [CrossRef]
34. Gunasekaran, S.; Govindan, S.; Ramani, P. Investigation of chemical and biological properties of an acidic polysaccharide fraction from *Pleurotus eous* (Berk.) Sacc. *Food Biosci.* **2021**, *42*, 101209. [CrossRef]
35. Włodarczyk, A.; Krakowska, A.; Sułkowska-Ziaja, K.; Suchanek, M.; Zięba, P.; Opoka, W.; Muszyńska, B. *Pleurotus* spp. *Mycelia Enriched in Magnesium and Zinc Salts as a Potential Functional Food. Molecules* **2021**, *26*, 162.
36. Arvidson, K.; Johasson, E.G. Galvanic current between dental alloys in vitro. *Scand. J. Dent. Res.* **1985**, *93*, 467–473. [CrossRef]
37. Neumann, M.; Goderska, K.; Grajek, K.; Grajek, W. Modele przewodu pokarmowego in vitro do badań nad biodostępnością składników odżywczych. *Żywność Nauka Technol. Jakość* **2006**, *1*, 30–45.
38. *Polish Pharmakopeia, Wydanie X*; PTFarm: Warszawa, Poland, 2014.
39. Opoka, W.; Muszyńska, B.; Rojowski, J.; Rumian, J. Gastroel–2014. Poland Patent Application P 417238, 18 May 2016.
40. Confortin, F.G.; Marchetto, R.; Bettin, F.; Camassola, M.; Salvador, M.; Dillon, A.J.P. Production of *Pleurotus sajor-caju* strain PS-2001 biomass in submerged culture. *J. Ind. Microbiol. Biotechnol.* **2018**, *35*, 1149. [CrossRef] [PubMed]
41. Rosado, F.R.; Germano, S.; Carbonero, E.R.; Costa, S.M.; Iacomini, M.; Kemmelmeier, C. Biomass and exopolysaccharide production in submerged cultures of *Pleurotus ostreatoroseus* Sing. and *Pleurotus ostreatus* "Xorida" (Jack.: Fr.) Kummer. *J. Basic Microbiol.* **2003**, *43*, 230–237. [CrossRef]
42. Poursaeid, N.; Azadbakht, A.; Balali, G.R. Improvement of zinc bioaccumulation and biomass yield in the mycelia and fruiting bodies of *Pleurotus florida* cultured on liquid media. *Biotechnol. Appl. Biochem.* **2015**, *175*, 3387–3396. [CrossRef] [PubMed]
43. Srikanth, M.; Sandeep, T.; Sucharitha, K.; Godi, S. Biodegradation of plastic polymers by fungi: A brief review. *Bioresour. Bioprocess.* **2022**, *9*, 42. [CrossRef]
44. Kała, K.; Krakowska, A.; Sułkowska-Ziaja, K.; Szewczyk, A.; Reczyński, W.; Opoka, W.; Muszyńska, B. Kinetics of extracted bioactive components from mushrooms in artificial digestive juices. *Int. J. Food Prop.* **2017**, *20*, 1796–1817. [CrossRef]

molecules

MDPI

Article

Screening Antioxidant Components in Different Parts of Dandelion Using Online Gradient Pressure Liquid Extraction Coupled with High-Performance Liquid Chromatography Antioxidant Analysis System and Molecular Simulations

Xia Cao [1,†], Gaoquan Li [1,†], Juying Xie [1], Mengqi Wu [2], Wenhao Wang [2], Li Xiao [1,*] and Zhengming Qian [1,2,*]

1 College of Medical Imaging Laboratory and Rehabilitation, Xiangnan University, Chenzhou 423000, China; seasonsyang@sina.cn (X.C.); gaogaokf2024@outlook.com (G.L.); xjy597189746@163.com (J.X.)
2 Key Laboratory of State Administration of Traditional Chinese Medicine, Dongguan HEC Cordyceps R&D Co., Ltd., Dongguan 523850, China
* Correspondence: 13762582461@163.com (L.X.); qianzhengming1982@126.com (Z.Q.)
† These authors contributed equally to this work.

Abstract: Utilizing online gradient pressure liquid extraction (OGPLE) coupled with a high-performance liquid chromatography antioxidant analysis system, we examined the antioxidative active components present in both the aerial parts and roots of dandelion. By optimizing the chromatographic conditions, we identified the ferric reducing–antioxidant power system as the most suitable for online antioxidant reactions in dandelion. Compared to offline ultrasonic extraction, the OGPLE method demonstrated superior efficiency in extracting chemical components with varying polarities from the samples. Liquid chromatography–mass spectrometry revealed twelve compounds within the dandelion samples, with nine demonstrating considerable antioxidant efficacy. Of these, the aerial parts and roots of dandelion contained nine and four antioxidant constituents, respectively. Additionally, molecular docking studies were carried out to investigate the interaction between these nine antioxidants and four proteins associated with oxidative stress (glutathione peroxidase, inducible nitric oxide synthase, superoxide dismutase, and xanthine oxidase). The nine antioxidant compounds displayed notable binding affinities below −5.0 kcal/mol with the selected proteins, suggesting potential receptor–ligand interactions. These findings contribute to enhancing our understanding of dandelion and provide a comprehensive methodology for screening the natural antioxidant components from herbs.

Keywords: dandelion; *Taraxacum mongolicum*; antioxidant; high-performance liquid chromatography; mass spectrometry; online gradient pressure liquid extraction; molecular docking

1. Introduction

Dandelion (*Taraxacum mongolicum* Hand. -Mazz.), belonging to the Asteraceae family, is extensively distributed across the Northern Hemisphere [1]. This non-toxic herbaceous plant has both edible and medicinal properties. Various parts of the dandelion, including its leaves, roots, and flowers, are utilized in a range of food products. The young leaves, for instance, are often consumed fresh in salads [2–7]. The flowers are used in brewing and dessert preparation [6,7]. Nutritional analysis reveals that dandelion is abundant in minerals, proteins, fibers, vitamins, and balanced trace elements, positioning it as a valuable source of micronutrients [2–5]. From a medicinal perspective, dandelion boasts a rich tradition in herbal medicine due to its potential benefits such as diuretic, heat relief and detoxification, choleretic, and anti-inflammatory [1,8]. Recently, the antioxidant activity of dandelion has attracted considerable attention. The plant is rich in antioxidant components, including polyphenols and flavonoids, making it a promising natural source for the development of antioxidants [9,10]. Despite the recognition of their health benefits and potential as a reservoir of antioxidant compounds, conventional methodologies to evaluate the antioxidant activity in dandelion are significantly resource-intensive and inefficient in terms of time. Traditional antioxidant assays require substantial resources and a significant time cost, which inherently limits the speed and efficiency of identifying active compounds. In response to the urgent need for a more effective approach, our study is focused on developing a quick and efficient method to screen potential antioxidant compounds in dandelion.

Traditional methods for assessing antioxidant activity, such as 2,2-azino-bis-3-ethylbenzothiazoline-6-sulphonic acid (ABTS), ferric reducing–antioxidant power (FRAP), and 2,2-diphenyl-1-picrylhydrazyl (DPPH) assays, involve reacting with the sample solution and observing the resultant products to determine antioxidant activity. For example, the FRAP method operates by reducing Fe^{3+}-ferric tripyridyltriazine (TPTZ) to Fe^{2+}-TPTZ under acidic conditions, facilitated by the antioxidant. The absorbance at a wavelength of 593 nm is then measured, serving as an indicator of the total antioxidant capacity of the sample [11].

Lately, integrating high-performance liquid chromatography (HPLC) with antioxidant assessment has become a refined approach for analyzing antioxidants, attributed to its distinct benefits. Offline coupling methods such as ABTS/DPPH-HPLC provide an approach for screening potential antioxidants [12,13]. However, the analysis process can be complex, involving steps such as sample extraction, reaction of the sample with ABTS/DPPH, and HPLC analysis of both the sample solution and the reaction solution. To address these limitations, recent developments have introduced online coupling methods of HPLC and antioxidant evaluation, such as HPLC-DPPH/ABTS/FRAP [14–16]. A significant advantage of these methods is their ability to integrate separation and activity evaluation. This integration can effectively bypass some unnecessary analytical steps, thereby saving time. These methods streamline the analysis process, making it more efficient and less labor-intensive, and could be applied in antioxidant research.

Currently, both offline and online HPLC coupling methods for antioxidant evaluation have become valuable tools in rapidly screening antioxidants from natural products. However, despite the elimination of some unnecessary separation steps, the overall process remains time-consuming and labor-intensive, particularly in sample pre-treatment. For the preparation of samples, methods including microwave-assisted procedures, the process of reflux, and sonication techniques are frequently employed in offline contexts. However, considerable quantities of organic solvents, along with extensive extraction periods or substantial sample volumes, are necessitated by such approaches. Additionally, thermal decomposition during the extraction process can decrease the activity of some natural antioxidants. Therefore, while current HPLC-antioxidant evaluation methods have advanced the field significantly, further optimization is necessary to improve efficiency and reduce labor intensity in areas such as sample preparation and extraction. Such improvements could enhance the speed and accuracy of antioxidant analysis in natural products. To address these challenges, an online gradient pressure liquid extraction (OGPLE) combined with an HPLC-based antioxidant assay system was established [17,18]. In this system, a hollow guard cartridge is filled with a powdered sample to form an online extraction cell. The cell is interfaced with an HPLC setup via a valve featuring six ports, utilizing the mobile phase for both extraction and subsequent analytical procedures. Utilizing this system, the extraction of milligram-level samples can be completed within a few minutes, and separation and detection can be achieved in a single run. Additionally, this system forgoes employing any solvents besides the mobile phase, offering a simple, rapid, and efficient green method that minimizes or even eliminates sample handling processes.

Molecular docking serves as a time-efficient and cost-effective computational approach utilized in forecasting the interactions of small molecules with their target proteins. This approach entails positioning a small molecule (referred to as a ligand) within a protein's binding domain (defined as a receptor), where ongoing modifications in spatial arrangement are conducted to predict the most favorable interaction point and manner of the ligand–receptor ensemble [19]. Molecular docking, a staple in drug discovery efforts, aids in identifying promising lead compounds, enhancing their properties, and forecasting interactions between ligands and proteins [20,21]. While its reliance on computer theoretical simulations often necessitates supplementary experimental validation, many researchers recognize it as a potent instrument for advancing research fields due to its efficiency and predictive capacity [22].

The current study presents the development of an OGPLE combined with an HPLC-based antioxidant assay system for screening the main antioxidant components in the dandelion aerial parts and dandelion roots. Also utilized was the technique of liquid chromatography–tandem mass spectrometry (LC-MS/MS), which pinpointed the compounds specific to antioxidant activity. Subsequently, to investigate the potential interactions between these compounds and oxidative-stress-related proteins, we selected four related proteins, inducible nitric oxide synthase (iNOS), glutathione peroxidase (GSH-Px), superoxide dismutase (SOD), and xanthine oxidase (XOD), for molecular docking analyses to determine their binding affinities. This study aims to elucidate the antioxidant active components of the aerial parts and roots of dandelion, and develop a comprehensive approach for screening antioxidant components from natural products.

2. Results and Discussion

2.1. Refining Experimental Parameters

2.1.1. Optimization of Chromatographic Conditions

Three chromatographic columns were evaluated in this study: an Agilent ZORBAX SB-AQ column (4.6 mm × 150 mm, 5 µm), a Thermo Hypersil GOLD aQ column (4.6 mm × 150 mm, 5 µm), and a Welch Ultimate AQ-C_{18} column (4.6 mm × 150 mm, 5 µm). Among them, the Thermo Hypersil GOLD aQ column demonstrated superior separation performance and peak shapes (Figure S1). Additionally, the experiment compared three online antioxidant systems: ABTS, FRAP, and DPPH (Figure S2). The results indicated that the FRAP system exhibited distinct antioxidant reaction peaks with a more stable baseline, making it more suitable for online antioxidant reaction systems. In contrast, the ABTS and DPPH systems experienced more significant baseline fluctuations, which are detrimental to the analysis of compound antioxidant activity.

2.1.2. Optimization of OGPLE

In this experiment, the types of dispersants and the mixing ratio of sample to dispersant were investigated. By comparing two different dispersants, diatomite and acid-washed diatomite, it was found that the use of acid-washed diatomite resulted in better chromatographic peak shapes and separation performance (Figure S3). The mixing ratios of sample to acid-washed diatomite (1:5, 1:10) were examined (Figure S4). The chromatograms obtained from the 1:5 for sample to acid-washed diatomite showed better response values. Therefore, a mixing ratio of 1:5 was chosen for the OGPLE process. Additionally, a second round of OGPLE was carried out on the samples. The results demonstrated that no chromatographic peaks were observed during the second extraction (Figure 1), suggesting that one OGPLE cycle could effectively obtain all substances present in the samples.

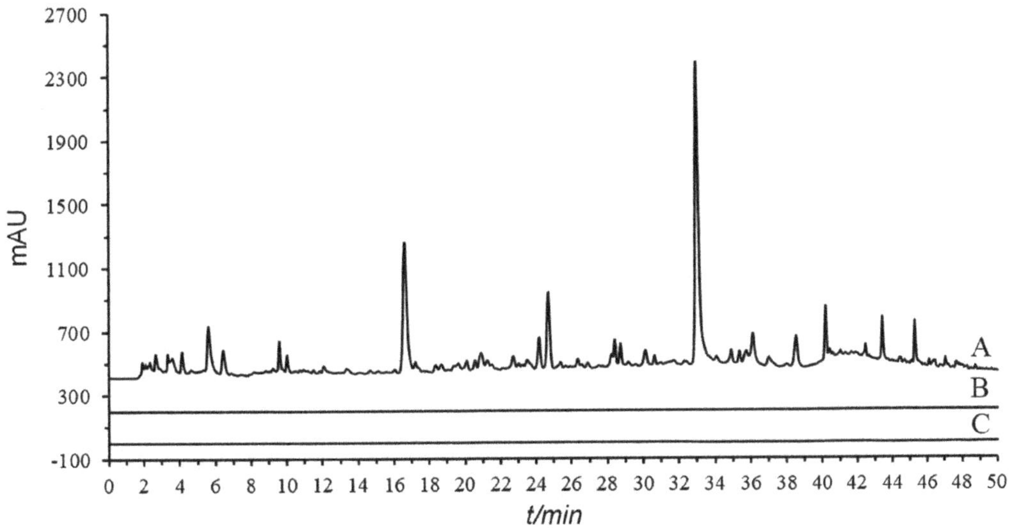

Figure 1. Chromatographic comparison of single and double OGPLE: first OGPLE (A), second OGPLE (B), and blank (C). OGPLE: online gradient pressure liquid extraction.

2.1.3. Comparison between OGPLE and Offline Extraction

During the experimental process, it was discovered that employing offline ultrasonic extraction methods proved challenging for the simultaneous extraction of chemical components with different polarities from dandelion. The experiments utilized three solvents—water, 50% methanol, and anhydrous methanol—for extracting the samples (Figure 2). The results indicated that certain low-polarity components in the extracts obtained with water and 50% methanol were missing (Figure 2b), while extracts obtained with anhydrous methanol were missing some high-polarity components (Figure 2a). This study employed the OGPLE method, which utilizes a gradient alteration in mobile phase polarity for sample extraction. This gradient shifted from high to low polarity, meeting the extraction conditions for compounds of different polarities present in the sample. This approach enabled the maximum extraction of components within a single analysis, significantly saving on analysis time and solvent consumption.

The stability of the aqueous ultrasound extract of dandelion was analyzed in 24 h. Both caffeic acid and cichoric acid exhibited a relative standard deviation (RSD) greater than 10%, with caffeic acid increasing by 38.1% and cichoric acid decreasing by 29.2% after 24 h. This indicates that the aqueous extract of dandelion is not stable. In contrast, the OGPLE method used in this study allows the compounds to be extracted by the mobile phase and immediately analyzed by HPLC. This reduces the decomposition and reaction of compounds, accurately reflecting the content of various components in the sample and enhancing the accuracy of the analysis.

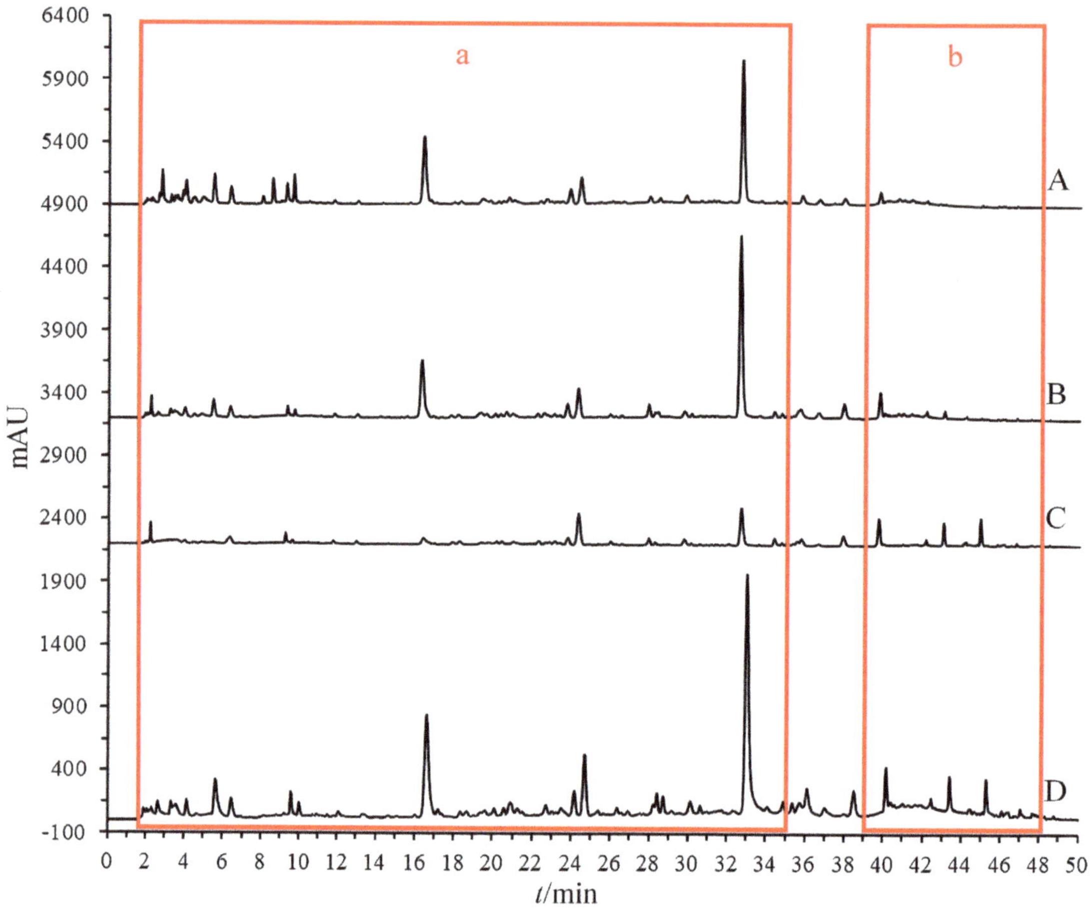

Figure 2. Chromatographic comparison of offline and online extraction methods (water extraction (A), 50% methanol extraction (B), anhydrous methanol extraction (C), and OGPLE (D)) with high-polarity components (a) and low-polarity components (b). OGPLE: online gradient pressure liquid extraction.

2.2. Method Validation

This experiment investigated the specificity of the method by analyzing blank solutions, standard solutions, and dandelion aerial part sample solutions (Figure S5). The method demonstrated good chromatographic peak separation with no interference from impurity peaks at the elution positions, allowing for the qualitative analysis of the samples. Furthermore, following the procedure outlined in the "OGPLE of Sample" Section, three dandelion aerial part samples were extracted online. The findings demonstrated that the RSD for the principal chromatographic peaks across all three samples remained below 10%, indicating that the established method has good repeatability.

2.3. Identification and Quantification of Antioxidant Active Components from Dandelion Aerial Parts and Dandelion Roots

In this investigation, OGPLE combined with the HPLC-based antioxidant assay system was employed for the separation and subsequent online antioxidant analysis of compounds in the dandelion aerial part and dandelion root samples (Figure 3). The analysis included measuring compound retention times at two distinct wavelengths: initially at 275 nm (Detector 1) to detected the compounds, followed by post-reaction detection at 593 nm

(Detector 2) to measure the absorbance of the Fe^{2+}-TPTZ complex. This phase enabled linking specific substances with their antioxidative actions, determined by analyzing the chromatographic peaks observed pre- and post-reaction. It was found that nine compounds demonstrated antioxidant activity in the dandelion aerial part samples (Figure 3(B2)) and four in the dandelion root samples (Figure 3(C2)).

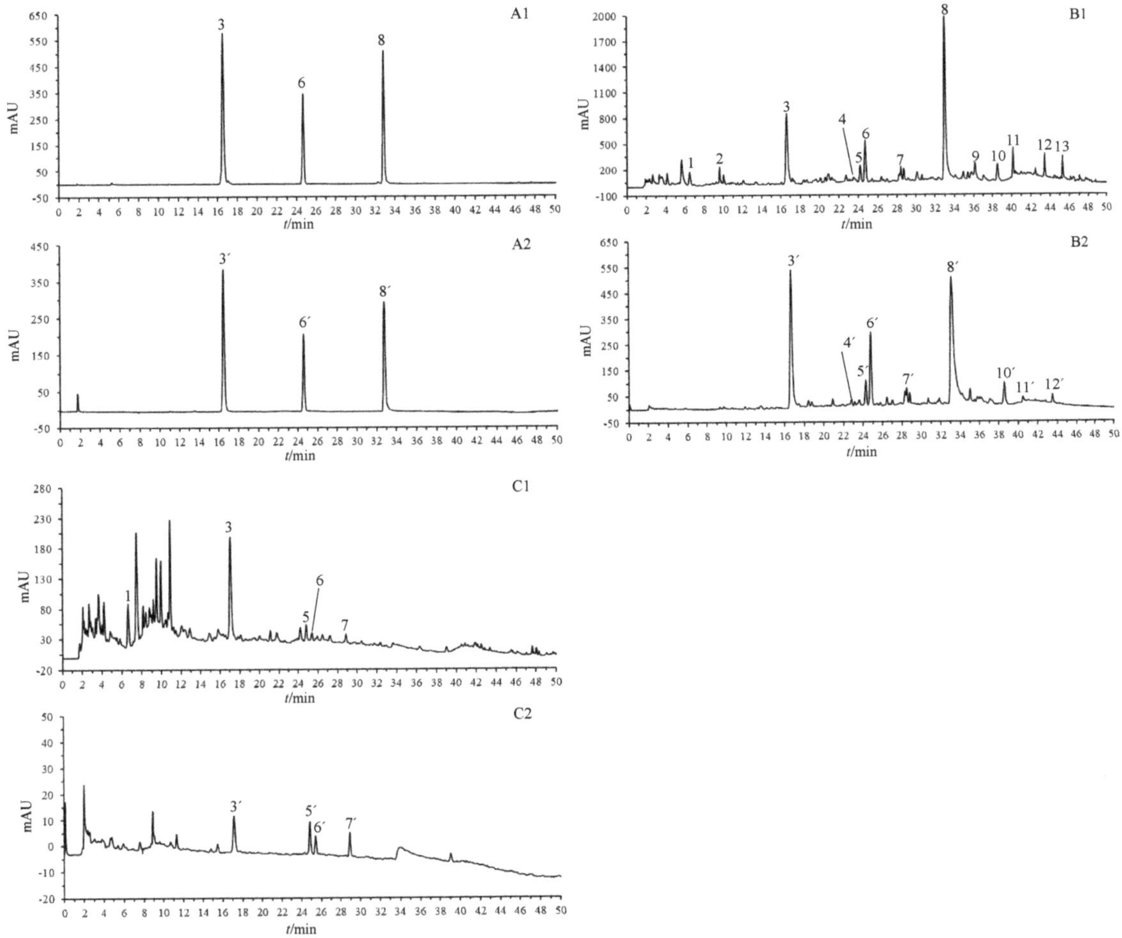

Figure 3. OGPLE coupled with HPLC antioxidant analysis system for dandelion aerial part and dandelion root samples. (**A**) Standard solution; (**B**) dandelion aerial parts; (**C**) dandelion roots. (**A1,B1,C1**) Total compound chromatograms. (**A2,B2,C2**) Fe^{2+}-TPTZ peaks detected after reaction with FRAP solution. Peak Assignments: 1. uric acid, 2. adenosine, 3. caftaric acid, 4. aesculetin, 5. neochlorogenic acid, 6. caffeic acid, 7. caffeoylmalic acid, 8. cichoric acid, 9. luteolin 7-O-glucoside, 10. 3,5-Di-caffeoylquinic acid, 11. phlorizin, 12. luteolin, 13. unknown. OGPLE: online gradient pressure liquid extraction; HPLC: high-performance liquid chromatography; TPTZ: ferric tripyridyltriazine; FRAP: ferric reducing–antioxidant power.

MS was used to identify compounds in the dandelion aerial part and root samples (Table 1). Twelve compounds were identified in the dandelion aerial part samples, namely uric acid (1), adenosine (2), caftaric acid (3), aesculetin (4), neochlorogenic acid (5), caffeic acid (6), caffeoylmalic acid (7), cichoric acid (8), luteolin 7-O-glucoside (9), 3,5-di-caffeoylquinic acid (10), phlorizin (11), and luteolin (12). In the dandelion root samples, five compounds were identified: uric acid (1), caftaric acid (3), neochlorogenic acid (5), caffeic acid (6), and caffeoylmalic acid (7). The identification was based on the retention times of the reference components, as well as MS data with known studies and databases [23–27].

Table 1. MS data of components in the aerial parts and roots of dandelion.

Peak No.	Compound	RT	Molecular Formula	Exact Mass	Adduct Ion	Precursor Ion	Product Ion	Reference
1	Uric acid	6.453	$C_5H_4N_4O_3$	168.0283	$[M-H]^-$	167.0200	124.0141, 96.0190	[23]
2	Adenosine	9.593	$C_{10}H_{13}O_4N_5$	267.0968	$[M+H]^+$	268.1042	136.0620	[24]
3	Caftaric acid	16.606	$C_{13}H_{12}O_9$	312.0481	$[M-H]^-$	311.0411	179.0343, 149.0083, 135.0442	[25]
4	Aesculetin	23.506	$C_9H_6O_4$	178.0266	$[M-H]^-$	177.0183	149.0236, 133.0284, 105.0334	[23]
5	Neochlorogenic acid	24.173	$C_{16}H_{18}O_9$	354.0951	$[M-H]^-$	353.0879	191.0555	[24]
6	Caffeic acid	24.693	$C_9H_8O_4$	180.0423	$[M-H]^-$	179.0340	135.0441, 133.0283	[26]
7	Caffeoylmalic acid	28.419	$C_{13}H_{12}O_8$	296.0532	$[M-H]^-$	295.0460	179.0343, 135.0441, 133.0132, 115.0025	[27]
8	Cichoric acid	33.006	$C_{22}H_{18}O_{12}$	474.0798	$[M-H]^-$	473.0728	311.0412, 293.0306, 179.0342, 149.0082, 135.0040	[26]
9	Luteolin 7-O-glucoside	36.132	$C_{21}H_{20}O_{11}$	448.1006	$[M-H]^-$	447.0940	285.0407	[26]
10	3,5-Di-caffeoylquinic acid	38.519	$C_{25}H_{24}O_{12}$	516.1268	$[M-H]^-$	515.1201	353.0886, 191.0556, 179.0343	[26]
11	Phlorizin	40.179	$C_{21}H_{24}O_{10}$	436.1370	$[M-H]^-$	435.1303	273.0775, 167.0342	[24]
12	Luteolin	43.425	$C_{15}H_{10}O_6$	286.0477	$[M-H]^-$	285.0405	175.0395, 151.0028, 133.0285	[23]
13	Unknown	45.279	-	-	$[M-H]^-$	329.2335	229.1445, 211.1338, 199.4364, 171.1018	-

RT: retention time; MS: mass spectrometry.

Within the dandelion aerial part samples, the compounds exhibiting antioxidant activity included caftaric acid (3), aesculetin (4), neochlorogenic acid (5), caffeic acid (6), caffeoylmalic acid (7), cichoric acid (8), 3,5-di-caffeoylquinic acid (10), phlorizin (11), and luteolin (12). For the dandelion root samples, the antioxidant active components were identified as caftaric acid (3), neochlorogenic acid (5), caffeic acid (6), and caffeoylmalic acid (7).

The quantitative analysis of nine antioxidant active compounds in the aerial parts and roots of the dandelion samples was carried out (Table S1). Caffeic acid content was measured by employing its calibration curve ($y = 63.85x - 30.831$; $R^2 = 0.9995$) of the standard. Concurrently, the concentrations of the other eight compounds were estimated by the caffeic acid calibration curve as a reference, owing to the absence of standards for these compounds. This surrogate approach, utilized in previous research, is a common method in the absence of specific standard substances [28]. The quantitative analysis of the dandelion revealed that caftaric acid (1.24 mg/g dry weight [DW], Percentage: 22%), caffeic acid (0.478 mg/g DW, Percentage: 8%), and cichoric acid (2.75 mg/g DW, Percentage: 48%) were the major antioxidants in the aerial parts, with caftaric acid (0.205 mg/g DW, Percentage: 78%) being the primary antioxidant in the roots. Previous research has indicated that caftaric acid and caffeic acid in plant extraction remain stable for over

70 days at room temperature [29], while cichoric acid is stable for more than 30 days in dry, light-protected conditions [30]. The addition of these antioxidant compounds into food products may prolong storage duration [31]. Consequently, it can be considered that the antioxidant active compounds in dandelion products are highly stable. In addition, the quantitative analysis results showed that the content of antioxidants in the aerial portion is higher than in the root. To confirm this finding, an offline antioxidant analysis was conducted, and the result indicates that the antioxidant ability of the dandelion aerial portions is more than 4 times that of the roots (Figure S6). This result is consistent with the quantitative analysis of antioxidants in dandelion. These findings align with previous studies [32], demonstrating the reliability of online antioxidant analysis as a valid method. It is important to acknowledge that the reported findings are preliminary, and a thorough evaluation of the antioxidant capacity of these identified compounds would necessitate detailed quantitative analyses along with functional assays. Future research endeavors should strive to accurately quantify these compounds in dandelion while investigating their synergistic effects and bioavailability. This study will help to improve our knowledge of the potential antioxidant efficacy of dandelion, and improve the quality evaluation and product development of dandelion.

2.4. Molecular Docking Analysis

In the field of antioxidant research, molecular docking analysis has been widely adopted, particularly for exploring how antioxidants interact with specific targets to elicit their functions [33,34]. This approach not only aids in understanding the molecular-level efficacy of antioxidants, but also lays a solid theoretical foundation for online antioxidant screening. Molecular docking analysis was conducted to explore the interaction dynamics of nine promising antioxidant compounds with four proteins associated with oxidative stress, including GSH-Px, iNOS, SOD, and XOD (Table 2). The docking conformations and key binding residues of dandelion's principal antioxidant compounds in the interaction with receptors were also simulated and delineated (Figure 4; Table 3). Typically, an affinity score below −5.0 kcal/mol indicates receptor–ligand coupling [35]. The results indicate that all nine antioxidant compounds exhibit affinity scores below −5.0 kcal/mol with the four proteins. GSH-Px, an enzyme capable of peroxide clearance [36], demonstrated affinity scores of −5.3 kcal/mol for caftaric acid, −5.4 kcal/mol for caffeic acid, and −6.2 kcal/mol for cichoric acid. iNOS, responsible for nitric oxide production and the modulation of inflammation and protective stress responses [37], demonstrated affinity scores of −6.8 kcal/mol for caftaric acid, −6.8 kcal/mol for caffeic acid, and −9.4 kcal/mol for cichoric acid. SOD, known for its antioxidant properties [38], demonstrated affinity scores of −7.1 kcal/mol for caftaric acid, −6.2 kcal/mol for caffeic acid, and −7.5 kcal/mol for cichoric acid. XOD, involved in generating reactive oxygen species [39], demonstrated affinity scores of −8.5 kcal/mol for caftaric acid, −7.0 kcal/mol for caffeic acid, and −8.8 kcal/mol for cichoric acid. These docking results highlight the significant binding affinities of the nine compounds with the selected receptors, offering insights into their antioxidant effects and other potential biological activities. Of course, these findings will need further validation through experimental studies. In conclusion, the molecular docking analysis revealed significant binding affinities of the nine compounds with the selected receptors, shedding light on their antioxidant effects and potential biological activities.

Table 2. Parameters and outcomes of molecular docking analysis.

Receptor	Number of Points	Center Grid Box	Spacing	Ligand	Affinity (kcal/mol)
GSH-Px	X-dimension = 104 Y-dimension = 96 Z-dimension = 126	X center = 12.253 Y center = 8.525 Z center = 14.137	0.375	Caftaric acid	−5.3
				Aesculetin	−5.4
				Neochlorogenic acid	−6.9
				Caffeic acid	−5.4
				Caffeoylmalic acid	−6.3
				Cichoric acid	−6.2
				3,5-Di-caffeoylquinic acid	−7.3
				Phlorizin	−5.9
				Luteolin	−6.8
iNOS	X-dimension = 82 Y-dimension = 102 Z-dimension = 126	X center = 124.075 Y center = 110.548 Z center = 61.561	0.914	Caftaric acid	−6.8
				Aesculetin	−7.5
				Neochlorogenic acid	−7.9
				Caffeic acid	−6.8
				Caffeoylmalic acid	−7.3
				Cichoric acid	−9.4
				3,5-Di-caffeoylquinic acid	−8.3
				Phlorizin	−9.1
				Luteolin	−9.5
SOD	X-dimension = 56 Y-dimension = 126 Z-dimension = 110	X center = 11.495 Y center = 36.951 Z center = 32.456	1.000	Caftaric acid	−7.1
				Aesculetin	−6.3
				Neochlorogenic acid	−8.1
				Caffeic acid	−6.2
				Caffeoylmalic acid	−7.0
				Cichoric acid	−7.5
				3,5-Di-caffeoylquinic acid	−7.9
				Phlorizin	−8.5
				Luteolin	−8.5
XOD	X-dimension = 88 Y-dimension = 92 Z-dimension = 72	X center = 23.556 Y center = 32.646 Z center = 101.417	1.000	Caftaric acid	−8.5
				Aesculetin	−7.3
				Neochlorogenic acid	−8.3
				Caffeic acid	−7.0
				Caffeoylmalic acid	−7.7
				Cichoric acid	−8.8
				3,5-Di-caffeoylquinic acid	−9.8
				Phlorizin	−9.2
				Luteolin	−9.8

iNOS: inducible nitric oxide synthase; GSH-Px: glutathione peroxidase; SOD: superoxide dismutase; XOD: xanthine oxidase.

Table 3. The key binding residues of three major antioxidants.

Receptor	Ligand	Binding Site Interactions	Key Residues in Interaction
GSH-Px	Caftaric acid	Hydrophobic interactions	TRP8A
		Hydrogen bonds	SER7A, TRP8A, LYS116A, SER120A, ARG121A
		π-stacking	TRP8A
	Caffeic acid	Hydrophobic interactions	TRP8A, LYS116A
		Hydrogen bonds	SER7A, TRP8A, LYS116A
	Cichoric acid	Hydrophobic interactions	TRP8A, LYS116A, SER120A
		Hydrogen bonds	SER7A, TRP8A, LYS119A, SER120A, HIS123A
iNOS	Caftaric acid	Hydrophobic interactions	ASP125B, LYS248B, ILE494B
		Hydrogen bonds	ILE494B
		Salt bridges	HIS493B
	Caffeic acid	Hydrophobic interactions	HIS493A
		Hydrogen bonds	ARG252A, THR492A, ILE494A
		Salt bridges	HIS493A
	Cichoric acid	Hydrophobic interactions	PRO344A, VAL346A
		Hydrogen bonds	GLN257A, ASN348A, GLY365A, TYR485A
		π-Stacking	PHE363A
SOD	Caftaric acid	Hydrophobic interactions	LEU150B
		Hydrogen bonds	ASN-1A, LEU105A, GLY107B, SER110A, SER110B, ILE112B, ARG114A, ARG114B
	Caffeic acid	Hydrophobic interactions	ILE112A, ILE112B
		Hydrogen bonds	ASN-1B, SER110A, SER110B, ILE112A, ARG114B
	Cichoric acid	Hydrophobic interactions	ILE112C, ILE112D
		Hydrogen bonds	GLY-3C, GLY-3D, ASN-1C, ASN-1D, ASN106D, SER110D, ILE112D
		Salt bridges	ARG114C
XOD	Caftaric acid	Hydrophobic interactions	ARG1222C
		Hydrogen bonds	ASN272B, ASP429B, ARG606C, ASN830C, ARG1222C
		Salt bridges	ARG606C
	Caffeic acid	Hydrophobic interactions	PHE604C
		Hydrogen bonds	ARG32A, ASP594C, LEU605C, ARG824C
	Cichoric acid	Hydrophobic interactions	LEU257B, GLU263B, ILE264B, ILE353B
		Hydrogen bonds	LYS256B, VAL259B, GLY260B, ASN261B, SER347B, GLY350B, ILE353B

iNOS: inducible nitric oxide synthase; GSH-Px: glutathione peroxidase; SOD: superoxide dismutase; XOD: xanthine oxidase.

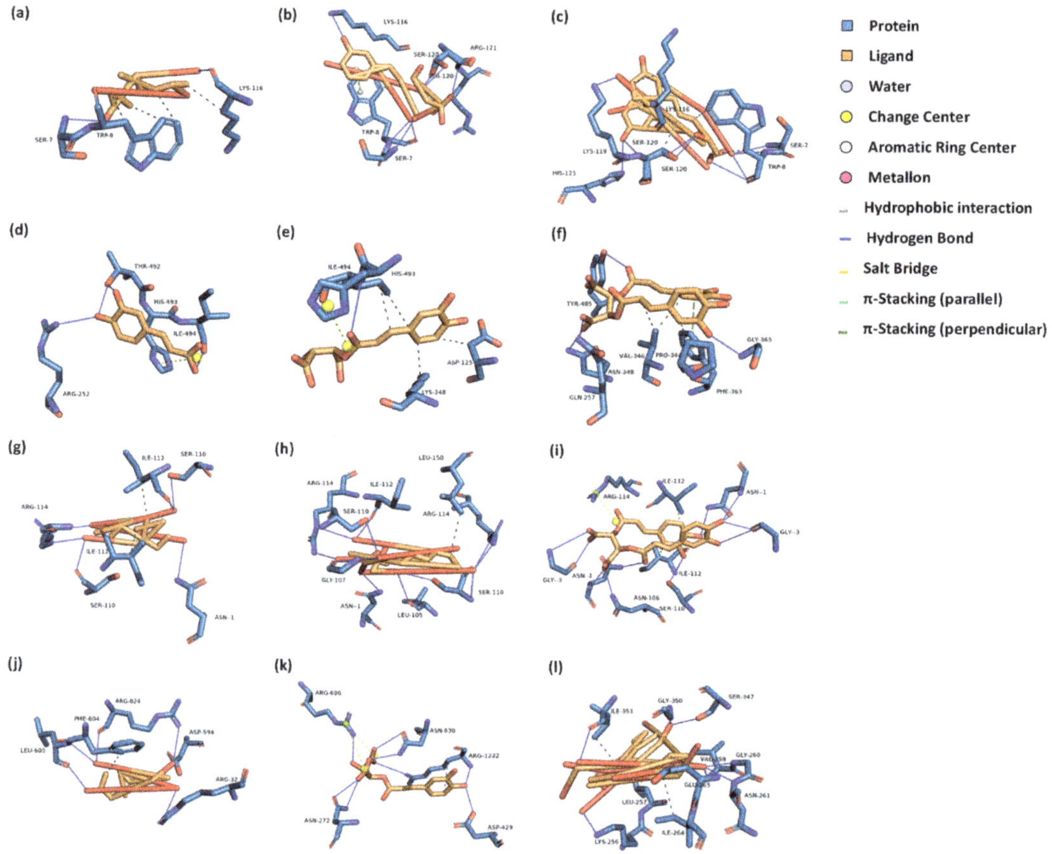

Figure 4. Docking conformations of dandelion's principal antioxidant compounds with GSH-Px, iNOS, SOD, and XOD receptors. (**a–c**) Interactions with the GSH-Px receptor for caffeic acid, caftaric acid, and cichoric acid, respectively. (**d–f**) Interactions with the iNOS receptor for caffeic acid, caftaric acid, and cichoric acid, respectively. (**g–i**) Interactions with the SOD receptor for caffeic acid, caftaric acid, and cichoric acid, respectively. (**j–l**) Interactions with the XOD receptor for caffeic acid, caftaric acid, and cichoric acid, respectively. GSH-Px: glutathione peroxidase; iNOS: inducible nitric oxide synthase; SOD: superoxide dismutase; XOD: xanthine oxidase.

3. Materials and Methods

3.1. Chemicals and Reagents

Ultrapure water was produced in a Milli-Q Advantage A10 water purification system (Merck KGaA., Darmstadt, Hesse, Germany). Acetic acid (HPLC grade), formic acid (HPLC grade), 2,2′-azino-bis(3-ethylbenzothiazoline-6-sulfonic acid) diammonium salt (98%), 2,4,6-tris(2-pyridyl)-s-triazine (99%), and potassium persulfate (Analytical Reagent) were obtained from Shanghai Aladdin Biochemical Technology Co., Ltd. (Shanghai, China). Ferric chloride (Chemically Pure) was obtained from Shanghai Macklin Biochemical Co., Ltd. (Shanghai, China). Sodium acetate anhydrous (Analytical Reagent) was obtained from Guangdong Guanghua Sci-Tech Co., Ltd. (Shantou, Guangdong, China). Celatom®, acid-washed and 2,2-diphenyl-1-picrylhydrazyl (100%) were obtained from Sigma-Aldrich, Inc. (Merck KGaA., Darmstadt, Hesse, Germany). Anhydrous methanol (Analytical Reagent) was obtained from Fuchen (Tianjin) Chemical Reagent Co., Ltd. (Tianjin, China). Acetonitrile (HPLC grade) was obtained from ANPEL Laboratory Technologies (Shanghai) Inc. (Shanghai, China). Ethanol (HPLC grade) was obtained from Krude Company, Inc.

(Los Angeles, CA, USA). Caffeic acid (98.7%) was obtained from Shanghai Standard Technology Co., Ltd. (Shanghai, China). Hydrochloric acid (Analytical Reagent), phosphorus pentoxide (Analytical Reagent), and sodium hydroxide (Analytical Reagent) were obtained from Chengdu Kelong Chemical Co., Ltd. (Chengdu, Sichuan, China). Caftaric acid (98.8%) and cichoric acid (91.0%) were prepared in our laboratory.

3.2. *Sample of Dandelion*

The dandelion aerial part samples were procured from Bozhou Jingwan Chinese MEDICINE FACTORY (Bozhou, Anhui, China), with the origin being Bozhou, Anhui, China. The dandelion root samples were obtained from KANGMEI (Bozhou) HUATUO International Chinese MEDICINE City Co., Ltd. (Bozhou, Anhui, China), with the origin being Changbai Mountain in Jilin, China.

3.3. *Standard Solution Preparation*

A precise measure of standard substances, specifically caffeic acid, caftaric acid, and cichoric acid, was utilized to concoct a combined standard solution at an optimal concentration with 50% methanol. Following preparation, this solution underwent filtration with a 0.22 μm organic membrane filter and was preserved at a refrigerated temperature of 4 °C.

3.4. *Preparation of Antioxidant Assay Reagents*

3.4.1. FRAP Solution

Preparation of ferric chloride solution (20 mmol/L): We weighed out 652.6 mg of ferric chloride and dissolved it in 200 mL of ultrapure water, mixing thoroughly.

Preparation of TPTZ solution (10 mmol/L): We weighed out 624.7 mg of 2,4,6-tris(2-pyridyl)-s-triazine and dissolved it in 200 mL of 40 mmol/L hydrochloric acid, mixing thoroughly.

Preparation of sodium acetate–acetic acid buffer (300 mmol/L): We weighed 1.8225 g of anhydrous sodium acetate, dissolved it in 1 L of ultrapure water, then add 16 mL of acetic acid, and mix. We adjusted the pH to 3.6 using 1 mol/L sodium hydroxide or hydrochloric acid.

We combined the aforementioned solutions in a ratio of 1:1:10 (ferric chloride solution/TPTZ solution/sodium acetate–acetic acid buffer).

3.4.2. ABTS Solution

We weighed out 194.1 mg of 2,2′-azino-bis (3-ethylbenzothiazoline-6-sulfonic acid) diammonium salt and 68.6 mg of potassium persulfate, and dissolved them in ultrapure water (100 mL) to prepare a mixed solution with concentrations of 3.5 mmol/L for 2,2′-azino-bis(3-ethylbenzothiazoline-6-sulfonic acid) diammonium salt and 2.5 mmol/L for potassium persulfate. The reaction took place in the dark at 4 °C for 12 h. Prior to usage, we diluted the sample with ethanol and adjusted the solution to an absorbance of 1.0 at 750 nm on a Cary 60 UV-Vis Spectrophotometer (Agilent Technologies, Inc., Santa Clara, CA, USA).

3.4.3. DPPH Solution

To prepare a DPPH solution of 0.2 mmol/L concentration, we measured 24.2 mg of 2,2-diphenyl-1-picrylhydrazyl, which was then completely dissolved in 300 mL of anhydrous methanol.

3.5. *Offline Extraction of Sample*

Referencing previous methodologies [40], we weighed approximately 0.5 g of the sample powder three times, and placed each portion into a separate 15 mL centrifuge tube. To each tube, we added 5 mL of ultrapure water, 50% methanol, and anhydrous methanol, respectively. The mixture was subjected to ultrasonic extraction for 30 min (at a power of 380 W and a frequency of 37 kHz), followed by thorough mixing. We collected the supernatant and filtered it through a 0.22 μm membrane. The filtrate obtained was used as the sample solution.

3.6. OGPLE of Sample

The OGPLE combined with the HPLC-based antioxidant assay system was constructed based on previous studies (Figure 5) [17,18]. The sample powder was combined with acid-washed diatomite at a 1:5 ratio. Subsequently, approximately 5.0 mg of the homogeneous mixture was measured accurately and loaded into an empty SecurityGuard Standard, and the void was filled with acid-washed diatomite and sealed securely at both ends with a 0.22 μm filter membrane. It was then loaded back into the SecurityGuard Cartridge (3.0 × 4.0 mm; Phenomenex, Inc., Torrance, CA, USA) to form a sample extraction cell. The assembled sample extraction cell and a blank extraction cell were connected to the LC system via a six-way valve, utilizing the mobile phase for the online micro-extraction process. When the sample was analyzed, the six-port valve transitioned to the sample extraction cell to facilitate online extraction. When replacing the sample became necessary, the valve was adjusted to the blank extraction cell, thereby efficiently purifying the system and preparing it for the introduction of a new sample.

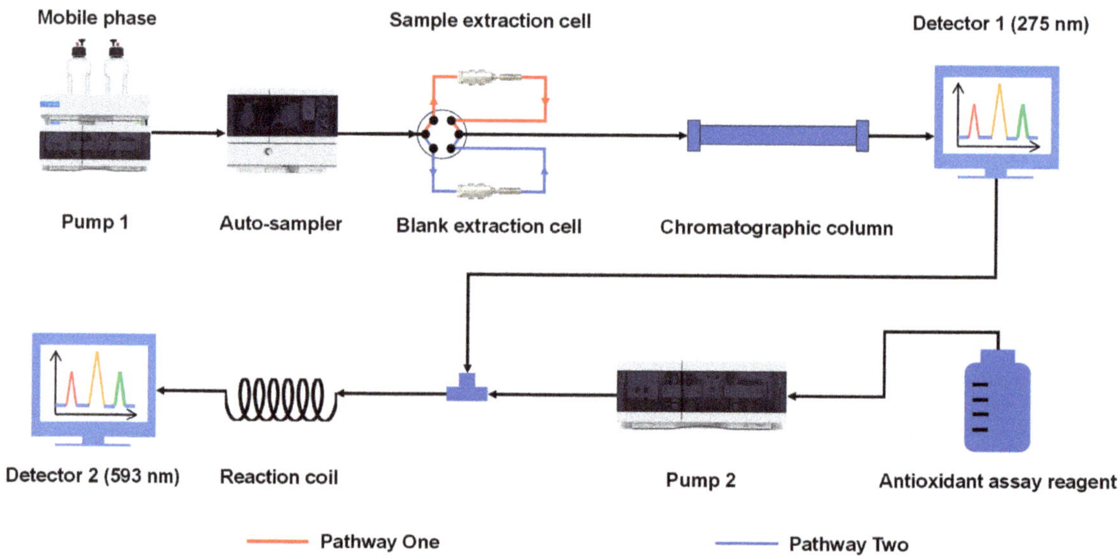

Figure 5. Schematic of the OGPLE-HPLC antioxidant analysis system. OGPLE: online gradient pressure liquid extraction; HPLC: high-performance liquid chromatography.

3.7. HPLC Instruments and Conditions

The chromatographic analysis in this study was carried out using an Agilent 1260 Infinity LC System (Agilent Technologies, Inc.). The chromatographic column employed was a Thermo Hypersil GOLD aQ column (4.6 mm × 150 mm, 5 μm) maintained at a temperature of 25 °C. The composition of the mobile phase was 0.1% formic acid (A) and acetonitrile (B), with the gradient elution described as follows: initiating with 0% B from 0 to 3 min; a linear escalation to 4% B from 3 to 7 min; a rise to 5% B between 7 and 11 min; an increment to 10% B from 11 to 20 min; a progressive increase to 18% B from 20 to 30 min; an elevation to 21% B between 30 and 36 min; culminating in a surge to 70% B from 36 to 50 min. The flow rate was set at 1 mL/min, with the detection wavelength at 275 nm. A volume of injection of 5 μL was used for offline extraction analysis without additional sample introduction for online micro-extraction analysis.

3.8. Offline Antioxidant Analysis

Measurements were conducted using a BioTek Synergy H1 Multimode Reader (Agilent Technologies, Inc.). Dandelion sample offline extraction solutions (50% methanol extraction) were diluted fivefold for aerial parts and twofold for roots. Concurrently, solutions of the caffeic acid standard were created, with concentrations measuring 19.8, 99.0, 198, 297, and 396 μg/mL. Assays were set up as follows: a blank group (200 μL FRAP solution + 10 μL blank solvent); a sample group (200 μL FRAP solution + 10 μL sample solution); and a reference compound group (200 μL FRAP solution + 10 μL caffeic acid standard solution of varying concentrations). These mixtures were gently combined in a 96-well plate, incubated at room temperature for 5 min, agitated on a plate shaker for 1 min, and then measured for absorbance at 593 nm. To quantify the antioxidative capacity of the samples, the absorbance readings from the sample extracts were calculated against the standard curve generated from the caffeic acid standards. The antioxidant power was expressed in terms of caffeic acid equivalents, providing a comparative measure of antioxidative strength across different samples.

3.9. Online Antioxidant Analysis

As depicted in Figure 5, after separation through HPLC, the sample solution was combined with antioxidant assay reagents. These reagents were administered by another HPLC pump at a delivery rate of 0.5 mL/min. The mixture then reacted within a 1.5 m × 0.25 μm Polyether ether ketone tube before entering the ultraviolet detector for the assessment of antioxidant components. These components were detected at wavelengths of 593 nm (FRAP), 750 nm (ABTS), and 517 nm (DPPH).

3.10. Preparation of Caftaric Acid and Cichoric Acid

The preparation of standard substances for caftaric acid and cichoric acid was carried out on the Agilent 1260 Infinity HPLC System (Agilent Technologies, Inc.). Sample extracts were separated on an Agilent ZORBAX SB-C_{18} column (9.4 mm × 250 mm, 5 μm) with an injection volume of 100 μL and a flow rate of 4 mL/min. The elution gradient and mobile phase followed the protocol described in the "HPLC Instruments and Conditions" Section. Eluates for caftaric acid (16 to 17 min) and cichoric acid (33 to 34 min) were collected across five consecutive injections, pooled separately for each compound, and then subjected to vacuum drying. The drying process involved placing Petri dishes containing phosphorus pentoxide along with another dish containing the eluates inside a vacuum oven (Model DZF6050; Gongyi Yuhua Instrument Co., Ltd., Gongyi, Henan, China), depressurizing for 30 min and then drying continuously at room temperature (25 ± 5 °C) for 15 h. The residues were then re-dissolved in 50% methanol to obtain caftaric acid crude extract solution and cichoric acid crude extract solution.

The extracted solutions underwent refinement using a Thermo Hypersil GOLD aQ column (4.6 mm × 150 mm, 5 μm), employing a mobile phase consisting of 0.1% formic acid (A) and acetonitrile (B). Specifically, for the caftaric acid raw extract, a gradient elution strategy was utilized that started with 4% B for 0 to 16 min; increased to 10% B from 16 to 18 min; and was then held constant at 10% B for 18 to 23 min. The procedure involved an injection volume of 80 μL and a flow rate set at 1 mL/min. The eluate collected between 13 and 14 min was vacuum-dried following the previous method, yielding the caftaric acid standard. Similarly, the cichoric acid crude extract solution was processed with an isocratic elution of 14% B for 20 min, with an injection volume of 80 μL and a flow rate of 1 mL/min. The eluate collected between 14 and 15 min was also vacuum-dried, resulting in the cichoric acid standard.

3.11. LC-MS/MS Qualitative Analysis Conditions

Chromatographic separation utilized the Vanquish Flex UHPLC System (Thermo Fisher Scientific Inc., Franklin, MA, USA), and mass spectrometric analysis was conducted using a Q Exactive Focus Orbitrap LC-MS/MS System (Thermo Fisher Scientific Inc.).

The LC conditions are described in the "HPLC Instruments and Conditions" Section. Performing in both the positive and negative ion states, the MS analysis used full scan and dd-MS^2 scanning modes. The comprehensive scan extended across a mass-to-charge (m/z) range of 100 to 1500, acquiring data at a resolution of 70,000. For dd-MS^2 scans, the scope was set from m/z 50 to 1000, with a resolution setting of 17,500. Normalized collision energies were programmed in a stepped sequence of 10, 20, and 40. The ionization method employed was heated electrospray ionization (H-ESI), characterized by sheath, auxiliary (Aux), and sweep gas flow rates of 60 arb, 20 arb, and 3 arb, respectively. The spray voltage was uniformly maintained at 3500 V across both positive and negative modes. Additionally, the capillary and Aux gas heaters were regulated at temperatures of 400 °C and 450 °C, respectively.

3.12. *Quantitative Analysis of Antioxidant Compounds*

Adhering to our previously described "Standard Solution Preparation" Section, caffeic acid standard solutions were prepared with concentrations of 1.98 μg/mL, 99.0 μg/mL, 198 μg/mL, 396 μg/mL, and 594 μg/mL. Samples underwent extraction per the method outlined in the "OGPLE of Sample" Section, and the HPLC analysis followed the protocols outlined in the "HPLC Instruments and Conditions" Section. The calculation of the content of all antioxidant compounds in the sample was based on the caffeic acid standard curve.

3.13. *Molecular Docking Study*

Using the PyMOL (https://pymol.org/) and AutoDock Vina 1.2.0 software tools, this study primarily conducted molecular docking [41–43]. This experimental process necessitates the three-dimensional structure of the ligand and the crystal structure of the receptor. To achieve this, the three-dimensional structures of the antioxidants were created utilizing the Chem3D version 20.0.0.41 software (PerkinElmer Inc., Shelton, CT, USA). The crystallographic configurations for pertinent proteins, such as GSH-Px (PDB ID: 2WGR), iNOS (PDB ID: 1M8D), SOD (PDB ID: 1TO4), and XOD (PDB ID: 1FIQ), were acquired from the RCSB Protein Data Bank [44]. Subsequently, the receptor protein underwent processing utilizing the AutoDock Vina and PyMOL software, which included dehydration, removal of the ligand, and the addition of charges. The AutoDock Vina software was then employed to frame the entire receptor protein by adjusting the coordinates, length, width, and height of the Grid Box, and the parameters were recorded accordingly. Ultimately, the AutoDock Vina software was used to conduct molecular docking of the receptor and ligand, generating a maximum of 100 models, and the affinity results were obtained. The concluding step involved analyzing and visualizing the docking results with the lowest binding energies using PyMOL, and the binding site residues were identified through the Protein–Ligand Interaction Profiler [45].

4. Conclusions

OGPLE combined with the HPLC-based antioxidant assay system established in this study significantly enhances the rapid identification of potentially health-beneficial compounds within dandelion aerial part and dandelion root samples. This approach specializes in rapidly detecting antioxidants, providing valuable information for further research. Additionally, the molecular docking evaluations demonstrated the binding affinity of these compounds with the chosen receptors, illuminating their prospective antioxidant properties and various biological activities. In summary, the combined approach of OGPLE and the HPLC-based antioxidant assay, complemented by molecular docking analysis, offers an effective framework for the quick identification and validation of antioxidants in plant-based materials. This approach serves as a valuable tool in antioxidant research, facilitating the use of natural compounds for health-related purposes.

Supplementary Materials: The following supporting information can be downloaded at: https://www.mdpi.com/article/10.3390/molecules29102315/s1, Figure S1: Comparative chromatograms: Thermo Hypersil GOLD aQ (4.6 mm × 150 mm, 5 μm) (A), Agilent ZORBAX SB-AQ (4.6 mm × 150 mm, 5 μm) (B), and Welch Ultimate AQ-C_{18} (4.6 mm × 150 mm, 5 μm) (C); Figure S2: Comparison of chromatograms from online antioxidant systems: ferric reducing–antioxidant power (FRAP) (A), 2,2-azino-bis-3-ethylbenzothiazoline-6-sulphonic acid (ABTS) (B), and 2,2-diphenyl-1-picrylhydrazyl (DPPH) (C); Figure S3: Chromatograms of different dispersants: acid-washed diatomite (A) and diatomite (B); Figure S4: Chromatograms of different ratios (dandelion aerial part to acid-washed diatomite); Figure S5: Chromatograms of dandelion aerial parts (A), standard solution (B), and blank solution (C). Figure S6: Offline antioxidant analysis of dandelion aerial parts and roots (n = 3; mean ± standard deviation). Aerial parts: 86.1 mg/g; roots: 18.3 mg/g. *** $p < 0.001$ versus the roots, evaluated using Student's t-test. Table S1: The contents of nine antioxidant compounds in the aerial parts and roots of dandelion (n = 3).

Author Contributions: Conceptualization, Z.Q. and L.X.; funding acquisition, Z.Q. and J.X.; investigation, W.W., M.W. and X.C.; methodology, M.W.; project administration, Z.Q. and J.X.; supervision, G.L. and Z.Q.; writing—original draft, W.W., G.L., and X.C.; writing—review and editing, X.C., G.L., J.X., M.W., W.W., L.X. and Z.Q.; resources, Z.Q. All authors have read and agreed to the published version of the manuscript.

Funding: This work was supported by the Hunan Provincial Natural Science Foundation of China (Project Number: 2024JJ7536, Principal Investigator: Jing Chen).

Institutional Review Board Statement: Not applicable.

Informed Consent Statement: Not applicable.

Data Availability Statement: All data, models, and code generated or used during this study are presented in the submitted article.

Conflicts of Interest: Authors Mengqi Wu, Wenhao Wang and Zhengming Qian were employed by the company Dongguan HEC Cordyceps R&D Co., Ltd. The remaining authors declare that the research was conducted in the absence of any commercial or financial relationships that could be construed as a potential conflict of interest.

References

1. Hu, C. *Taraxacum*: Phytochemistry and health benefits. *Chin. Herb. Med.* **2018**, *10*, 353–361. [CrossRef]
2. Escudero, N.L.; De Arellano, M.L.; Fernández, S.; Albarracín, G.; Mucciarelli, S. *Taraxacum officinale* as a food source. *Plant Foods Hum. Nutr.* **2003**, *58*, 1–10. [CrossRef]
3. Souci, S.W.; Fachmann, W.; Kraut, H. *Food Composition and Nutrition Tables*, 7th ed.; Med Pharm Scientific: Stuttgart, Germany, 2008.
4. European Commission Scientific Committee on Food. *Opinion of the Scientific Committee on Food on the Revision of Reference Values for Nutrition Labeling*; European Commission: Brussels, Belgium, 2003.
5. Shi, S.; Zhao, Y.; Zhou, H.; Zhang, Y.; Jiang, X.; Huang, K. Identification of antioxidants from *Taraxacum mongolicum* by high-performance liquid chromatography–diode array detection–radical-scavenging detection–electrospray ionization mass spectrometry and nuclear magnetic resonance experiments. *J. Chromatogr. A* **2008**, *1209*, 145–152. [CrossRef] [PubMed]
6. Leung, A.Y.; Foster, S. *Encyclopedia of Common Natural Ingredients Used in Food, Drugs and Cosmetics*, 2nd ed.; John Wiley & Sons, Inc.: New York, NY, USA, 1996.
7. Olas, B. New perspectives on the effect of dandelion, its food products and other preparations on the cardiovascular system and its diseases. *Nutrients* **2022**, *14*, 1350. [CrossRef] [PubMed]
8. Bisset, N.G.; Wichtl, M. *Herbal Drugs and Phytopharmaceuticals: A Handbook for Practice on a Scientific Basis*; CRC Press: Boca Raton, FL, USA, 1994; pp. 486–489.
9. Wang, R.; Li, W.; Fang, C.; Zheng, X.; Liu, C.; Huang, Q. Extraction and identification of new flavonoid compounds in dandelion *Taraxacum mongolicum Hand.-Mazz.* with evaluation of antioxidant activities. *Sci. Rep.* **2023**, *13*, 2166. [CrossRef] [PubMed]
10. Grauso, L.; Emrick, S.; de Falco, B.; Lanzotti, V.; Bonanomi, G. Common dandelion: A review of its botanical, phytochemical and pharmacological profiles. *Phytochem. Rev.* **2019**, *18*, 1115–1132. [CrossRef]
11. Benzie, I.F.; Strain, J.J. The ferric reducing ability of plasma (FRAP) as a measure of "antioxidant power": The FRAP assay. *Anal. Biochem.* **1996**, *239*, 70–76. [CrossRef] [PubMed]
12. Kim, S.B.; Hwang, S.H.; Wang, Z.; Yu, J.M.; Lim, S.S. Rapid identification and isolation of inhibitors of rat lens aldose reductase and antioxidant in *Maackia amurensis*. *BioMed Res. Int.* **2017**, *2017*, 4941825. [CrossRef]

13. Liu, M.; Li, X.; Liu, Q.; Xie, S.; Zhu, F.; Chen, X. Preparative isolation and purification of 12 main antioxidants from the roots of *Polygonum multiflorum* Thunb. using high-speed countercurrent chromatography and preparative HPLC guided by 1, 1′-diphenyl-2-picrylhydrazyl-HPLC. *J. Sep. Sci.* **2020**, *43*, 1415–1422. [CrossRef]
14. Burnaz, N.A.; Küçük, M.; Akar, Z. An on-line HPLC system for detection of antioxidant compounds in some plant extracts by comparing three different methods. *J. Chromatogr. B* **2017**, *1052*, 66–72. [CrossRef]
15. Liu, C.; Lei, Y.; Dang, J.; Wang, W.; Zhang, J.; Mei, L.; Liu, Z.; Tao, Y.; Shao, Y. Preparative isolation of 1, 1-diphenyl-2-picrylhydrazyl inhibitors from *Ribes himalense* using medium-pressure and two-dimensional reversed-phase/reversed-phase liquid chromatography guided by an online HPLC-1, 1-diphenyl-2-picrylhydrazyl assay. *J. Sep. Sci.* **2021**, *44*, 1345–1352. [CrossRef] [PubMed]
16. Tian, S.; Yu, Y.; Liu, Q.; Guo, H.; Yu, J.; Wang, X.; Zhao, H. An integrated strategy for the geographical origin traceability of Goji berries by antioxidants characteristic fingerprint based online ultra-performance liquid chromatography-2, 2-diphenyl-1-picrylhydrazyl-photodiode array detector-mass spectrometry combined with multivariate statistics analysis. *J. Sep. Sci.* **2023**, *46*, 2200826.
17. Qian, Z.M.; Fang, B.W.; Chen, H.M.; Li, C.H.; Huang, Q.; Chen, L.; Li, W.J.; Li, D.Q. Online liquid microextraction coupled with HPLC-ABTS for rapid screening of natural antioxidants: Case study of three different teas. *J. Chromatogr. Sci.* **2020**, *58*, 875–879. [CrossRef] [PubMed]
18. Yang, W.Q.; Huang, Q.; Wu, M.Q.; Mei, Q.X.; Zou, Y.S.; Qian, Z.M.; Tang, D. Rapid screening and evaluation of natural antioxidants from leaf, stem, and root of Artemisia argyi by online liquid microextraction combined with HPLC-based antioxidant assay system coupled with calibration quantitative analysis. *J. Sep. Sci.* **2024**, *47*, 2300616. [CrossRef] [PubMed]
19. Pinzi, L.; Rastelli, G. Molecular Docking: Shifting Paradigms in Drug Discovery. *Int. J. Mol. Sci.* **2019**, *20*, 4331. [CrossRef] [PubMed]
20. Tajammal, A.; Siddiqa, A.; Irfan, A.; Azam, M.; Hafeez, H.; Munawar, M.A.; Basra, M.A.R. Antioxidant, molecular docking and computational investigation of new flavonoid. *J. Mol. Struct.* **2022**, *1254*, 132189. [CrossRef]
21. Singh, R.; Poke, A.V.; Ghosh, P.; Ganeshpurkar, A.; Swetha, R.; Singh, S.K.; Kumar, A. Pharmacophore-based virtual screening, molecular docking and molecular dynamics simulations study for the identification of LIM kinase-1 inhibitors. *J. Biomol. Struct. Dyn.* **2023**, *41*, 6089–6103. [CrossRef] [PubMed]
22. Crampon, K.; Giorkallos, A.; Deldossi, M.; Baud, S.; Steffenel, L.A. Machine-learning methods for ligand–protein molecular docking. *Drug Discov. Today* **2022**, *27*, 151–164. [CrossRef] [PubMed]
23. MassBank of North America. Available online: https://mona.fiehnlab.ucdavis.edu/ (accessed on 22 February 2024).
24. PubChem. Available online: https://pubchem.ncbi.nlm.nih.gov/ (accessed on 22 February 2024).
25. Khoza, B.S.; Gbashi, S.; Steenkamp, P.A.; Njobeh, P.B.; Madala, N.E. Identification of hydroxylcinnamoyl tartaric acid esters in Bidens pilosa by UPLC-tandem mass spectrometry. *S. Afr. J. Bot.* **2016**, *103*, 95–100. [CrossRef]
26. Schütz, K.; Kammerer, D.R.; Carle, R.; Schieber, A. Characterization of phenolic acids and flavonoids in dandelion (*Taraxacum officinale* WEB. ex WIGG.) root and herb by high-performance liquid chromatography/electrospray ionization mass spectrometry. *Rapid Commun. Mass Spectrom. Int. J. Devoted Rapid Dissem. Up Minute Res. Mass Spectrom.* **2005**, *19*, 179–186. [CrossRef]
27. The Human Metabolome Database. Available online: https://hmdb.ca/ (accessed on 22 February 2024).
28. Qian, Z.M.; Guan, J.; Yang, F.Q.; Li, S.P. Identification and quantification of free radical scavengers in Pu-erh tea by HPLC-DAD-MS coupled online with 2,2′-azinobis (3-ethylbenzthiazolinesulfonic acid) diammonium salt assay. *J. Agric. Food Chem.* **2008**, *56*, 11187–11191. [CrossRef] [PubMed]
29. Galmarini, M.V.; Maury, C.; Mehinagic, E.; Sanchez, V.; Baeza, R.I.; Mignot, S.; Zamora, M.C.; Chirife, J. Stability of individual phenolic compounds and antioxidant activity during storage of a red wine powder. *Food Bioprocess Technol.* **2013**, *6*, 3585–3595. [CrossRef]
30. Wills, R.B.H.; Stuart, D.L. Effect of handling and storage on alkylamides and cichoric acid in *Echinacea purpurea*. *J. Sci. Food Agric.* **2000**, *80*, 1402–1406. [CrossRef]
31. Bergeron, C.; Gafner, S.; Batcha, L.L.; Angerhofer, C.K. Stabilization of caffeic acid derivatives in *Echinacea purpurea* L. glycerin extract. *J. Agric. Food Chem.* **2002**, *50*, 3967–3970. [CrossRef] [PubMed]
32. Duan, L.; Zhang, C.; Zhao, Y.; Chang, Y.; Guo, L. Comparison of bioactive phenolic compounds and antioxidant activities of different parts of *Taraxacum mongolicum*. *Molecules* **2020**, *25*, 3260. [CrossRef] [PubMed]
33. Liu, C.; Lei, Y.; Liu, Y.; Guo, J.; Chen, X.; Tang, Y.; Dang, J.; Wu, M. An Integrated Strategy for Investigating Antioxidants from *Ribes himalense* Royle Ex Decne and Their Potential Target Proteins. *Antioxidants* **2023**, *12*, 835. [CrossRef] [PubMed]
34. Chen, J.; Huang, Q.; He, Z.; Tan, G.; Zou, Y.; Xie, J.; Qian, Z. Screening of Tyrosinase, Xanthine Oxidase, and α-Glucosidase Inhibitors from Polygoni Cuspidati Rhizoma et Radix by Ultrafiltration and HPLC Analysis. *Molecules* **2023**, *28*, 4170. [CrossRef] [PubMed]
35. Lu, J.; Song, H.P.; Li, P.; Zhou, P.; Dong, X.; Chen, J. Screening of direct thrombin inhibitors from Radix *Salviae miltiorrhizae* by a peak fractionation approach. *J. Pharm. Biomed. Anal.* **2015**, *109*, 85–90. [CrossRef] [PubMed]
36. Margis, R.; Dunand, C.; Teixeira, F.K.; Margis-Pinheiro, M. Glutathione peroxidase family—An evolutionary overview. *FEBS J.* **2008**, *275*, 3959–3970. [CrossRef]
37. Hemmrich, K.; Suschek, C.V.; Lerzynski, G.; Kolb-Bachofen, V. iNOS activity is essential for endothelial stress gene expression protecting against oxidative damage. *J. Appl. Physiol.* **2003**, *95*, 1937–1946. [CrossRef]

38. Bafana, A.; Dutt, S.; Kumar, A.; Kumar, S.; Ahuja, P.S. The basic and applied aspects of superoxide dismutase. *J. Mol. Catal. B Enzym.* **2011**, *68*, 129–138. [CrossRef]
39. Ardan, T.; Kovačeva, J.; Čejková, J. Comparative histochemical and immunohistochemical study on xanthine oxidoreductase/xanthine oxidase in mammalian corneal epithelium. *Acta Histochem.* **2004**, *106*, 69–75. [CrossRef] [PubMed]
40. Qian, Z.M.; Cheng, X.J.; Wang, Q.; Huang, Q.; Jin, L.L.; Ma, Y.F.; Xie, J.S.; Li, D.Q. On-line pre-column FRAP-based antioxidant reaction coupled with HPLC-DAD-TOF/MS for rapid screening of natural antioxidants from different parts of *Polygonum viviparum*. *RSC Adv.* **2023**, *13*, 9585–9594. [CrossRef] [PubMed]
41. Trott, O.; Olson, A.J. AutoDock Vina: Improving the speed and accuracy of docking with a new scoring function, efficient optimization, and multithreading. *J. Comput. Chem.* **2010**, *31*, 455–461. [CrossRef] [PubMed]
42. Eberhardt, J.; Santos-Martins, D.; Tillack, A.F.; Forli, S. AutoDock Vina 1.2.0: New docking methods, expanded force field, and python bindings. *J. Chem. Inf. Model.* **2021**, *61*, 3891–3898. [CrossRef]
43. Seeliger, D.; de Groot, B.L. Ligand docking and binding site analysis with PyMOL and Autodock/Vina. *J. Comput. Aided Mol. Des.* **2010**, *24*, 417–422. [CrossRef]
44. RCSB Protein Data Bank. Available online: https://www.rcsb.org/ (accessed on 22 February 2024).
45. Protein Ligand Interaction Profiler. Available online: https://plip-tool.biotec.tu-dresden.de/plip-web/plip/index (accessed on 22 February 2024).

molecules

MDPI

Article

Biochemical Assessments of Six Species of Edible Coastal Algae Collected from Tabuk Region in Saudi Arabia

Hala M. Bayomy [1,2,*] and Eman S. Alamri [1]

[1] Food Science and Nutrition Department, Science Faculty, University of Tabuk, Tabuk 71491, Saudi Arabia; ialamri@ut.edu.sa

[2] Food and Dairy Science and Technology Department, Faculty of Agriculture, Damanhour University, Damanhour 22516, Egypt

* Correspondence: hm.mohamed@ut.edu.sa

Abstract: In the first study focusing on the Red Sea's Tabuk coast, six edible species of the most common algae were collected to evaluate their approximate composition using AOAC methods, amino acids using ion-exchange chromatography, minerals using atomic absorption spectroscopy, phenolic compounds using the Folin–Ciocalteu method, and ferric-reducing antioxidant power. All the data were significantly ($p < 0.05$) different among all the studied species. The data indicated that the protein content ranged from 9.25% for *A. nodosum* to 20.06% for *H. musciformis*. *C. racemosa* had the highest lipid content of 7.57%. Phosphors varied from 68.2 mg/100 g for *A. nodosum* to 406 mg/100 g for *D. simplex*. The largest amounts of calcium (2458 mg/100 g) and iron (29.79 mg/100 g) were found in *C. racemosa*. The total essential amino acids ranged between 38.16 and 46.82% for *A. nodosum* and *D. simplex*, respectively. *F. vesiculosus* had the maximum content of phenolic compounds (11.06 mg GAE/g). *A. nodosum* had the highest antioxidant capacity (1.78 mg TE/g). The research concluded that algae are the main effort toward sustainable agriculture to meet the world's food needs. that algae may be used to improve food naturally. To satisfy the criteria for sustainable food, which is one of the pillars of NEOM, numerous studies are required to investigate the natural products available in the Red Sea.

Keywords: edible algae; chemical composition; amino acids; oxidoreductive; Saudi Arabia

1. Introduction

In recent years, the Red Sea has attracted attention due to its biological diversity and the abundance of algae biomass [1]. The type and distribution of the algae in the Red Sea are contingent upon various factors, including temperature and salinity changes, the depth zone, and the season [2,3]. Gomez-Zavaglia et al. [3] classified algae into microalgae and macroalgae. It should be noted that not all scientists agree on the division of algae, but the perspective is continually changing. One of the criteria for the classification of macroalgae is their pigmentation [4], which allows us to identify several large groups: red algae *Rhodophyta* (more than 6000 species), brown algae *Phaeophyceae* (more than 2000 species), and green algae chlorophyta (more than 1200 species).

Green marine macroalgae possess highly diverse forms, such as single or multicellular. The plastids are stained green by chlorophyll a and b, linked to carotenoids and xanthophyll. *Caulerpa racemosa* and *Ulva lactuca* are edible green algae [5–7]. Brown algae possess a multicellular structure. Fucoxanthin is a highly prevalent pigment found in brown algae, outweighing other pigments such as chlorophyll-a, chlorophyll-c, β-carotene, and other xanthophylls [8]. Brown algae are a type of marine macroalgae; they are plant species found in the Red Sea and in coastal regions all over the world [9]. Some species, including *Ascophyllum nodosum* and *Fucus vesiculosus*, are eaten in Asian countries in traditional recipes such as sushi [10]. Red marine macroalgae are multicellular, and a few are single-cellular. Their color is due to the presence of pink plastids, in which the red pigment, phycoerythrin, is bound to several other pigments, including chlorophyll. *Digenea simplex* and *Hypnea musciformis* are red edible algae [11].

Scientists have noted that marine macroalgae can be considered a valuable food resource due to their antioxidant activity, which is beneficial for human health [12]. It prevents the formation of free radicals, which are responsible for oxidative stress and cause multiple diseases [13,14]. Several studies have confirmed the positive correlation between free radicals and cancer, circulatory diseases, aging, rheumatic arthritis, and nervous system diseases [7,15].

Several authors have found that algae contain a high level of vital nutrients, including proteins, amino acids, polyunsaturated fatty acids, and polysaccharides, in addition to dietary fiber, vitamins, and minerals [7,8,13]. Algae has long been considered a staple in the Asian diet, adding to its high nutritional value [7,8,13]. In the last decade, numerous species of sea algae have been shown to contain various types of antioxidant substances—for example, sulfated polysaccharides, catechins, sterols, proteins, phlorotanins, and carotenoid pigments such as fucoxanthin and astaxanthin [16]. In addition, the polyphenols that are available in algae show great potential and possess stronger antioxidant activity than terrestrial plants [7,17,18]. Some researchers have found a correlation between algae's antioxidant activity and total phenolic compound levels [19]. The availability of such compounds in algae can protect the body from several illnesses and delay the aging process [20].

The chemical composition varies between algae species, which can be attributed to several factors, including salinity, temperature, location, light, seasonal period, and storage conditions [21,22]. Furthermore, algae are a great source of protein, which varies between 5% and 47% of the dry basic [7]. Recently, algae have been viewed as a cheap and innovative protein source with high nutritional quality [23].

The Red Sea is considered one of the most important regions of biological diversity on Earth and has a wide variety of algae species, but the nutritional benefits and chemical composition of the Red Sea algae are poorly understood. Therefore, the current study presents the first published data on the approximate composition, amino acids, minerals, phenolic compounds, and oxidoreductive compounds in common edible algae found on the Tabuk coast of Saudi Arabia. The investigation aims to identify the most prevalent edible algae on the Tabuk coast of the Red Sea and to estimate their nutritional composition for use in complementary studies in the near future.

2. Results

2.1. Proximate Chemical Composition

The calculated protein, lipid, ash, fiber, and carbohydrate percentages were derived from an ANOVA and are presented in Table 1 on a dry basis (db). Significant differences ($p < 0.05$) were found between all the species under study, which were composed of a wide variety of nutrients. The red algae *H. musciformis* had the highest protein content (20.06%, db), and the lowest value (9.25%, db) was found in the brown algae *A. nodosum*. The lipid content in the present study ranged from 0.93 to 7.57% for *D. simplex* and *C. racemosa*, respectively. The ash and crude fiber contents were higher among the algae species under study. The ash content varied from 13.64% in *C. racemosa* to 29.38% for *A. nodosum*, while the crude fiber varied from 11.75 to 34.81% for *U. lactuca* and *H. musciformis*, respectively. *H. musciformis* had the lowest carbohydrate content (23.56%), whereas *U. lactuca* had the highest content (54.52%).

2.2. Minerals

The mineral content of the six algae under study is given in Table 2. The values illustrated significant differences ($p < 0.05$) among all species. The highest calcium value (2458 mg/100 g) was found in the *C. racemosa* species, whereas the lowest value (476 mg/100 g) was found in *D. simplex*. On the other hand, the phosphorus content ranged between 68.29 and 747 mg/100 g for *U. lactuca* and *C. racemosa*, respectively. The iron content varied from 16.85 to 29.79 mg/100 g for *A. nodosum* and *C. racemosa*, respectively. The highest potassium content (7496 mg/100 g) was found in *D. simplex*, while the lowest content (477 mg/100 g) was found in *H. musciformis*. The sodium content ranged from 406 to 6156 mg/100 g for *H. musciformis* and *U. lactuca*, respectively.

Table 1. Proximate chemical composition (% *w*/*w*, db) of dried edible algae collected from Tabuk coast.

Type	Species	Protein	Lipid	Ash	Crude Fiber	Carbohydrate
Brown algae	*Ascophyllum nodosum*	9.25 ± 0.62 [d]	3.96 ± 0.07 [c]	29.38 ± 0.44 [a]	24.26 ± 0.53 [bc]	33.15 [cd]
	Fucus vesiculosus	13.71 ± 0.17 [cd]	3.74 ± 0.01 [c]	21.04 ± 0.31 [b]	19.61 ± 0.17 [c]	41.9 [bc]
Green algae	*Caulerpa racemosa*	18.39 ± 0.72 [ab]	7.57 ± 0.38 [a]	13.64 ± 0.16 [c]	13.82 ± 0.08 [d]	46.58 [ab]
	Ulva lactuca	14.62 ± 0.47 [cd]	5.29 ± 0.13 [b]	13.82 ± 0.06 [c]	11.75 ± 0.14 [d]	54.52 [a]
Red algae	*Digenea simplex*	15.17 ± 0.24 [bc]	0.93 ± 0.03 [d]	23.62 ± 0.77 [ab]	32.42 ± 1.21 [ab]	27.86 [cd]
	Hypnea musciformis	20.06 ± 0.53 [a]	1.83 ± 0.08 [d]	19.74 ± 0.52 [b]	34.81 ± 1.05 [a]	23.56 [d]

Mean ± standard deviation. Different letters in each column indicate statistically significant differences ($p < 0.05$).

Table 2. Mineral content in mg/100 g of dried edible algae collected from Tabuk coast.

Type	Species	Ca	P	Fe	K	Na
Brown algae	*Ascophyllum nodosum*	1026 ± 31.3 [b]	185 ± 28.5 [c]	16.85 ± 0.16 [c]	3643 ± 84.8 [b]	3895 ± 204 [b]
	Fucus vesiculosus	1049 ± 27.1 [b]	208 ± 9.3 [c]	20.66 ± 2.04 [b]	3758 ± 61.8 [b]	1869 ± 64.7 [d]
Green algae	*Caulerpa racemosa*	2458 ± 39.7 [a]	747 ± 13.6 [a]	29.79 ± 1.94 [a]	2873 ± 68.3 [c]	2188 ± 39.3 [c]
	Ulva lactuca	2393 ± 3.42 [a]	68.29 ± 1.6 [d]	25.63 ± 0.13 [a]	481 ± 23.7 [d]	406 ± 14.8 [f]
Red algae	*Digenea simplex*	476 ± 22.4 [c]	406 ± 17.2 [b]	18.03 ± 0.79 [bc]	7496 ± 142.7 [a]	1098 ± 15.7 [e]
	Hypnea musciformis	647 ± 8.2 [c]	371 ± 24.5 [b]	20.3 ± 0.86 [b]	477 ± 54.8 [d]	6156 ± 173.6 [a]

Mean ± standard deviation. Different letters in each column indicate statistically significant differences ($p < 0.05$).

As shown in Figure 1, *D. simplex* had the lowest Na/K ratio (0.15), while *H. musciformis* presented the highest Na/K ratio (12.90).

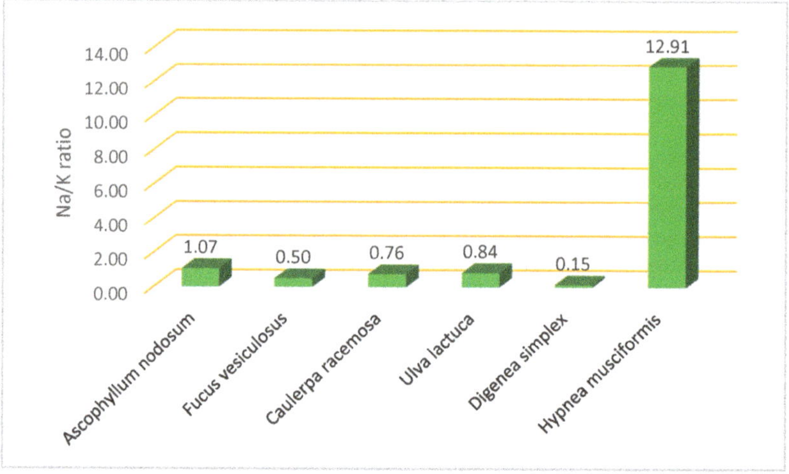

Figure 1. Na/K ratio in selected edible algae collected from Tabuk coast.

2.3. Total Phenolic Content and Oxidoreductive Compounds

Figure 2 shows the total phenolic content and oxidoreductive compounds of the six algae species taken from the Tabuk coast. The one-way ANOVA analysis showed that the oxidoreductive compounds differed significantly among the algae species under study. The brown species *F. vesiculosus* (11.06 mg GAE/g) and *A. nodosum* (9.38 mg GAE/g) had the highest levels of total phenolic compounds, followed by green algae, then the red algae *D. simplex* (0.72 mg GAE/g). *H. musciformis* (1.28 mg GAE/g) had the lowest level of total phenolic compounds. Among the tested algae, brown algae showed the highest reducing power (1.78 and 1.57 mg TE/g) for *A. nodosum* and *F. vesiculosus*, respectively, followed by green algae, and the lowest reducing power was observed in red algae *D. simplex* (0.58 mg TE/g) and *H. musciformis* (0.61 mg TE/g).

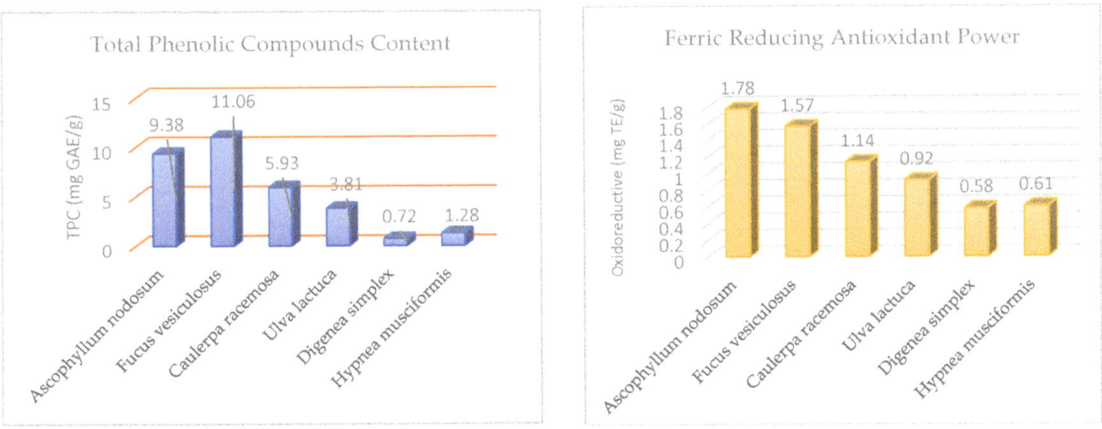

Figure 2. Total phenolic compounds (mg GAE/g) and ferric-reducing antioxidant power (mg TE/g) in selected edible algae.

2.4. Amino Acids

The amino acid composition (g/100 g protein) for the selected edible algae is shown in Table 3. The total essential amino acids ranged from 38.16 to 46.82 g/100 g protein for *A. nodosum* (which had the highest level of non-essential amino acids, 61.64 g/100 g protein) and *H. musciformis* (which had the lowest level of non-essential amino acids, 52.98 g/100 g protein), respectively. The calculated EAA/total AA ratio varied from 38.24% in *A. nodosum* to 46.91% in *H. musciformis*.

Table 3. Amino acid profile (g/100 g protein, db) for selected algae under study.

Amino Acids	***Ascophyllum nodosum***	***Fucus vesiculosus***	***Caulerpa racemosa***	***Ulva lactuca***	***Digenea simplex***	***Hypnea musciformis***	**Standard Error**
Aspartic acid	9.35 ^{d}	12.29 ^{b}	10.37 ^{c}	13.06 ^{a}	10.49 ^{c}	12.02 ^{b}	0.646
Serine	5.11 ^{c}	6.74 ^{b}	5.35 ^{c}	7.72 ^{a}	5.36 ^{c}	6.32 ^{b}	0.255
Glutamic acid	18.63 ^{a}	14.41 ^{b}	12.29 ^{c}	11.5 ^{c}	10.33 ^{d}	12.19 ^{c}	0.803
Proline	4.34 ^{d}	5.83 ^{a}	4.56 ^{c}	3.37 ^{f}	4.7 ^{b}	4.18 ^{e}	0.051
Glycine	6.53 ^{a}	4.75 ^{d}	6.34 ^{b}	5.83 ^{c}	5.76 ^{c}	4.31 ^{e}	0.171
Alanine	8.12 ^{a}	7.19 ^{b}	7.96 ^{a}	6.98 ^{b}	5.76 ^{c}	5.04 ^{d}	0.222
Arginine	4.55 ^{d}	5.07 ^{c}	6.83 ^{b}	7.31 ^{a}	4.19 ^{e}	5.41 ^{c}	0.341
Cysteine	0.24 ^{e}	1.53 ^{a}	0.39 ^{d}	0.87 ^{c}	1.04 ^{b}	1.05 ^{b}	0.045

Table 3. *Cont.*

Amino Acids	*Ascophyllum nodosum*	*Fucus vesiculosus*	*Caulerpa racemosa*	*Ulva lactuca*	*Digenea simplex*	*Hypnea musciformis*	Standard Error
Tyrosine	4.77 [b]	3.39 [e]	3.75 [c]	3.52 [d]	8.62 [a]	2.46 [f]	0.098
Total Non-EAA	61.64	61.2	57.84	60.16	56.25	52.98	0.327
Methionine	2.61 a	1.59 [c]	1.6 [c]	1.56 [c]	0.79 [d]	2.14 [b]	0.130
Isoleucine	4.21 [c]	3.71 [d]	6.17 [a]	3.86 [d]	4.7 [b]	6.18 [a]	0.029
Leucine	5.84 [d]	6.79 [b]	7.04 [a]	5.92 [d]	6.42 [c]	7.64 [a]	0.140
Phenylalanine	3.70 [e]	3.95 [d]	4.86 [b]	4.94 [b]	6.09 [a]	4.5 [c]	0.093
Histidine	4.37 [a]	3.92 [b]	2.48 [d]	2.55 [d]	3.11 [c]	4.31 [a]	0.080
Valine	5.84 [a]	4.76 [e]	4.69 [f]	5.42 [b]	5.03 [d]	5.23 [c]	0.011
Lysine	3.88 [f]	5.84 [c]	6.41 [b]	4.83 [e]	9.21 [a]	5.36 [d]	0.037
Tryptophan	1.76 [e]	2.31 [c]	2.07 [d]	1.76 [e]	2.52 [b]	2.99 [a]	0.046
Threonine	5.95 [d]	5.47 [f]	6.73 [c]	8.83 [a]	5.63 [e]	8.47 [b]	0.103
Total EAA	38.16	38.34	42.05	39.67	43.5	46.82	0.720
EAA/Total AA ratio	38.24	38.52	42.10	39.74	43.61	46.91	

Values are means of triplicates. Different letters in each row indicate statistically significant differences ($p < 0.05$).

3. Discussion

Ścieszka and Klewicka [24] note that algae are found throughout the world, covering two-thirds of all water bodies. Algae have been employed in a variety of food products due to their high concentrations of prebiotic and bioactive components. Agar, alginate, and carrageenan have been developed as a result of their gelling, thickening, and stabilizing qualities [7]. Furthermore, algae are utilized in food products as a dietary supplement and as an ingredient in functional foods, as well as to produce fermented foods. In addition, algae are used to improve the quality of meat products such as pasties, steaks, frankfurters, and sausages, as well as fish, fish products, and oils. Algae are also used to fortify cereal-based foods, including pasta, wheat, and bread. This was confirmed in [25], which summarized that algae is considered a real gold mine for many bioactive compounds, including protein, which is present in large proportions (about 30% to 55–60%), and therefore, it can be used to treat malnutrition, as more than 821 million people are malnourished globally due to inadequate protein-rich diets, resulting in increased demand for total dietary protein [25].

Babich et al. pointed out that the protein content varies greatly between different algae groups. Among the macroalgae, red and green algae (e.g., *P. vulgaris* and *U. lactuca*) often contain high levels of protein, in contrast to the low levels in most brown algae [26]. Fleurence [22] found that the protein content of most brown algae (*A. nodosum* and *F. vesiculosus*) used in industry is less than 15%, db. The protein content in certain green algae species, such as those in the genus *Ulva*, can range from 10% to 26%, db. However, several investigated red algae have been found to have higher protein levels for *Porphyra tenera* (47%, db) and *Palmaria palmata* (19.94%, db), as illustrated in [27,28]. The findings of the chemical composition analysis clearly show that the macronutrient content of algae varies. In our study, the highest content of protein and lipids was detected in the green algae *C. racemosa*. The highest content of protein and fiber, with a low content of lipids and carbohydrates, was found in *H. musciformis*. This indicates their great potential for use in the diet.

Our results are in line with those reported in [22,29,30]. According to [30], the proximate compositions of *C. racemosa* discovered on Martin Island were found as follows: protein, 19.72%; crude lipids, 7.65%; carbohydrates, 48.97%; fiber, 11.51%; and ash, 12.15. Moreover, in Lorenzo et al. [31], the protein, lipid, and ash content were recorded to be

8.70, 3.62, and 30.89%, db, respectively, for brown algae *A. nodosum* from the Spanish coast. Meanwhile, *F. vesiculosus* contained 12.99, 3.75, and 20.7% db protein, lipid, and ash content, respectively. The protein content of algae varies across the year, according to [2,32], with the highest levels in the winter and early spring and the lowest levels in the summer and early autumn. Our findings are in agreement with [33], who concluded that the chemical composition varies according to the species, geographical area, season, and estimation process. The chemical composition showed the high ash content of algae, indicating their high content of minerals. It was also found that all algae species in the current study contained many times more calcium than milk [26], indicating that they may be an excellent source of calcium for osteoporosis prevention and treatment, for developing children, and for pre- and postmenopausal women. The Ca content in all species under study was higher than that reported in [31] for *F. vesiculosus* (1160.27 mg/100 g). The analysis of the iron content indicated that the algae had greater iron content than many well-known dietary sources of iron, such as leafy green vegetables, legumes, nuts, and common cereals, which all contain between 2 and 4 mg/100 g [34]. One strategy to prevent iron deficiency, one of the most prevalent nutritional deficits around the world, could be to use these algae as a natural food source. The present results are in contrast to those of [34], which found that *A. nodosum* had 10 mg Fe/100 g, which was less than our result (16.85 mg/100 g), but *F. vesiculosus* and *U. lactuca* had 29 and 180 mg Fe/100 g, respectively, which were greater than our findings. All species of algae considered in the present study are good sources of potassium, especially *D. simplex* (7496 mg K/100 g, db), *F. vesiculosus* (3758 mg/100 g, db), and *A. nodosum* (3643 mg/100 g, db). The most prevalent element in seaweeds was K (3781.35–9316.28 mg/100 g db), followed by Na (1836.82–4575.71 mg/100 g db) and Ca (984.73–1160.27 mg/100 g db), according to [31]. Similar conclusions were reached in [35,36].

A low Na/K ratio is beneficial for human health as it reduces the risk of high blood pressure as well as cardiovascular disease [37]. The lowest Na/K ratio was found in *D. simplex*, so this species can be used as a flavorful alternative to table salt (NaCl) for those with hypertension and cardiovascular diseases. The same trend regarding the polyphenol content was found among all six species when studying the oxidoreductive compounds via the ferric-reducing antioxidant power assay.

Brown algae are the richest in polyphenols and show the greatest ability to scavenge free radicals (Figure 3), compared to the two other families of marine algae (green algae and red algae). This result is correlated with the findings of [38,39]. Phlorotannins represent the major phenolic compounds in marine brown algae [40]. It is known that phlorotannins have a variety of biological properties, such as the suppression of antiplasmin, heavy metal detoxification, antimicrobial activity, UV protection, and chemoprevention against vascular risk factors [17,41]. The authors of [42,43] demonstrated that the antioxidant activity was correlated with the total phenolics. Meanwhile, in [44], it was found that the oxidoreductive qualities of marine algae may arise from several bioactive compounds, including polyunsaturated fatty acids, especially omega 3, as well as pigments such as chlorophylls and carotenoids, vitamins, vitamin precursors, sulfated polysaccharides, and phenolic compounds, which are believed to be the most active elements responsible for marine algae's antioxidant functions. Therefore, marine algae are a good source of both water- and fat-soluble antioxidants [7,45]. These aspects indicate that the naturally occurring antioxidant substances present in edible algae can both shield food items from oxidative deterioration and prevent and/or treat illnesses brought about by free radicals. As a result, it is important to evaluate how the most common cooking techniques (boiled, steamed, and pancake) affect the levels of carotenoids and chlorophylls, total phenolic compounds, and antioxidant capacity. This represents an important direction in our future research. The highest levels of non-EAA were recorded for aspartic acid, glutamic acid, and alanine in every algae species under study, in line with the results found in [31]. According to [46], the glutamic and aspartic acid contents are influenced by the specific flavor and taste of seaweed. The green alga *U. lactuca* from Norway [47] had an EAA% of 40.30–40.79, which is closely related to our study (39.67%).

The calculated EAA/total AA ratio for the present species was similar to that found in [48] for different brown and red Spanish edible seaweeds. This ratio reached 42.72%, 40.82%, and 36.87% for *Undaria pinnatifida*, *Halomonas elongata*, and *Porphyra umbilicalis*, respectively. The EAA/total AA ratio for the studied species, however, was lower than the figures provided in [35], who noted ratios over 55% in *Gracilaria changii*. Leucine, threonine, and lysine were the most abundant essential amino acids found in the studied algae, and our findings are in line with those recorded by Lorenzo et al. [31] for *F. vesiculosus* and *A. nodosum*. However, they are in contrast to those of [35], who considered arginine to be the most abundant EAA in *G. changii*. In all of the algae species analyzed, high-quality protein was found, suggesting that they could be used as a supplement for human nutrition. Most algae tend to provide adequate amounts of total essential amino acids within the required limits for food because of their high concentrations of essential amino acids, particularly lysine, which is a limited amino acid in many foods.

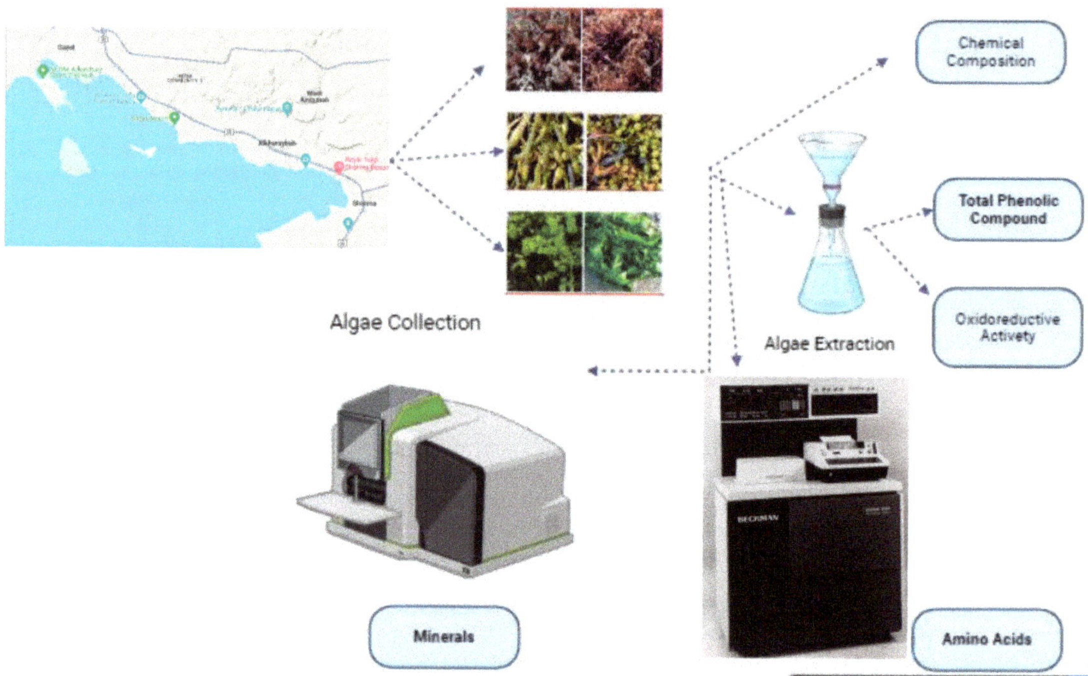

Figure 3. Flowchart to summarize materials and methods used.

4. Materials and Methods

Materials and methods can be summarized in the following flowchart (Figure 3).

4.1. Materials

All employed chemicals, reagents, and solvents were of the highest purity and were purchased from the Sigma-Aldrich Company (St. Louis, MI, USA).

Sampling Description

Algae samples were collected during September 2020 from three selected coastal zones in the region of Tabuk, including Sharma (27°55′48.4320″ N and 35°16′38.3808″ E), Alkhuraybah (28°03′24.0″ N 35°09′50.9″ E), and Gayal (28°07′33.5″ N 35°01′40.1″ E), as marked in Figure 4. According to Ansari and Ghanem [2], summer is the best season for the collection of algae samples, reflecting the diversity and density of Red Sea algae. Six species of edible algae were selected, including *Ascophyllum nodosum* and *Fucus vesiculosus* as brown algae, *Caulerpa racemosa* and *Ulva lactuca* as green algae, and *Digenea simplex* and *Hypnea musciformis* as red algae (Figure 5), and we collected approximately 3 kg of each species from each zone. All the species were identified with the assistance of the employees of the Oceanography Department at Alexandria University. To eliminate any associated contaminants, water was used to wash the obtained samples; then, they were rinsed with sterile water and dried in a shady environment at ambient temperature to prevent photolysis and thermal degradation. Then, they were minced well; the minced samples were preserved in airtight glass jars under freezing and then brought to the laboratory in iced conditions.

Figure 4. Selected coastal areas in the region of Tabuk (source: https://www.google.com/maps/@28.0766633,35.0502921,12z/data=!5m1!1e4?hl=en accessed on 22 January 2024).

Figure 5. (**a**) *Ascophyllum nodosum*, (**b**) *Fucus vesiculosu*, (**c**) *Caulerpa racemosa*, (**d**) *Ulva lactuca*, (**e**) *Digenea simplex*, and (**f**) *Hypnea musciformis*.

4.2. Methods

Proximate Chemical Composition: The proximate chemical composition of the selected algae was analyzed using the methodology of AOAC No. 930.15 for moisture, AOAC

984.13 for crude protein, AOAC 2003.05 for lipids, AOAC 978.10 for crude fiber, and AOAC 942.05 for ash. Results are presented as a percentage on a dry basis. The carbohydrate content (nitrogen-free extract) was computed by subtracting the total percentage of crude protein, fat, crude fiber, and ash content from one hundred [49].

Minerals: The concentrations of Fe, Ca, P, Na, and K were determined after transferring white to gray ash using treatment with 6N hydrochloric acid (HCl), followed by injection using atomic absorption spectroscopy (AAS), in line with the method explained in [50].

Amino Acids: First, 5 g of the sample and 2.5 mL of 6N hydrochloric acid were poured into a hermetically sealed hydrolysis tube. The assembly was then brought to a temperature of 110 °C. After 72 h, the hydrolysis was complete, and we determined the amino acids using ion-exchange chromatography (Beckman 7300 High-Performance Amino Acid Analyzer, Inc., Palo Alto, CA, USA). The alkaline hydrolysis of a sample was used to determine the tryptophan content [25].

4.2.1. Extraction Process for Algae Samples

The method recommended by Hemalatha et al. [51] was adopted to obtain the extracts, with some modifications. First, 2 g of the freeze-dried powder of different macroalgae was soaked in 40 mL of methanol for 24 h at ambient temperature. Twice more, extraction was performed, with the extracts filtered using Whatmann No. 1 filter paper (Sigma-Aldrich, St. Louis, MO, USA) each time, and the supernatants were kept at 4 °C to perform the analysis of total phenolics and oxidoreductive activity.

4.2.2. Total Phenolic Compounds

The total phenolic content in the selected algae was estimated using the Folin–Ciocalteu method, as explained by Dang et al. [52]. A spectrophotometer was used to detect the absorbance at 725 nm in comparison to a solvent blank. The total phenolic content was estimated using a calibration curve generated with gallic acid concentrations ranging from 4 M to 0.5 mM and represented as (mg GAE/g).

4.2.3. Determination of Ferric Reducing Antioxidant Power

The ferric-reducing antioxidant power (FRAP) assay was used. This procedure is based on the reduction of a ferric-tripyridyl triazine complex to its ferrous-colored state after incubation at 37 °C for 10 min in the presence of antioxidants. The absorbance of the reaction mixture at 593 nm was determined using a spectrophotometer. Calibration was performed using a series of concentrations between 200 and 1000 μM of $FeSO_4$ $7H_2O$ to plot the standard curve. FRAP was expressed as mg TE/g [53].

4.2.4. Statistical Analysis

The obtained results were recorded as the average of three replicates ± SD (standard deviation) except the data of amino acid, with the exception of amino acids, using Duncan's multiple range test with a one-way analysis of variance (ANOVA) at the significance level of $p < 0.05$. SPSS (version 21.0) was used.

5. Conclusions

Among all types of edible algae collected from Tabuk on the Red Sea coast, a high content of protein (18.39%), lipids (7.57%), ash (13.64%), crude fiber (13.83%), calcium (2458 mg/100 g), phosphorus (747 mg/100 g), and iron (29.79 mg/100 g) was observed in green algae *C. racemosa*, while a high content of protein (20.06%) and fiber (34.81%) with a low content of lipids (1.83%) and carbohydrates (23.56%) was found in *H. musciformis*. These possess great potential for use in the diet. All species of algae considered in the present study are good sources of potassium, especially *D. simplex*, which had the lowest Na/K ratio, while *H. musciformis* presented the highest Na/K ratio. Therefore, *D. simplex* can be used as a flavorful alternative to table salt (NaCl) for those with hypertension and cardiovascular diseases. The brown species *F. vesiculosus* and *A. nodosum* had the highest levels of total phenolic compounds and oxidoreductive activity, as determined using the ferric-reducing antioxidant power assay, followed by green algae and red algae. The natural antioxidant compounds found in edible algae can protect food products from oxidative degradation, as well as prevent and/or treat diseases caused by free radicals. Leucine, threonine, and lysine were the most abundant essential amino acids found in the algae under study. In all the algae species analyzed, high-quality protein was also found, suggesting that they could be used as a supplement in human nutrition. Moreover, most algae tend to provide adequate amounts of total essential amino acids within the required limits for food because of their high concentrations of essential amino acids, which ranged between 38.16 and 46.82% for *A. nodosum* and *D. simplex*, respectively. *F. vesiculosus* had the maximum content of phenolic compounds (11.06 mg GAE/g). *A. nodosum* had the highest antioxidant capacity (1.78 mg TE/g). The study concludes that the nutrient content of algae varies and that the Red Sea algae can be used as a natural food supplement that is rich in many nutrients. However, there is a need for additional studies to explore the natural products of the Red Sea and their potential to fulfill the requirements of healthy food, which is one of the pillars of the NEOM national project.

Author Contributions: Conceptualization, H.M.B. and E.S.A.; methodology, H.M.B. and E.S.A.; software, H.M.B. and E.S.A.; validation, H.M.B. and E.S.A.; formal analysis, H.M.B. and E.S.A.; investigation, H.M.B. and E.S.A.; resources, H.M.B. and E.S.A.; data curation, H.M.B. and E.S.A.; writing—original draft preparation, H.M.B. and E.S.A.; writing—review and editing, H.M.B. and E.S.A.; visualization, H.M.B. and E.S.A.; supervision, H.M.B. and E.S.A.; project administration, H.M.B. and E.S.A. All authors have read and agreed to the published version of the manuscript.

Funding: This research received no external funding.

Institutional Review Board Statement: Not applicable.

Informed Consent Statement: Not applicable.

Data Availability Statement: Data are contained within the article.

Acknowledgments: We thank the staff of the Department of Oceanography at Alexandria University for identifying the algae species.

Conflicts of Interest: The authors declare no conflicts of interest.

References

1. Sonnewald, M.; El-Sherbiny, M.M. Editorial: Red Sea Biodiversity. *Mar. Biodivers.* **2017**, *47*, 991–993. [CrossRef]
2. Ansari, A.A.; Ghanem, S.M. Seasonal Variation in the Growth Responses of Some Chlorophytic Algal Flora of the Red Sea. *Egypt. J. Aquat. Res.* **2017**, *43*, 129–134. [CrossRef]
3. Gomez-Zavaglia, A.; Prieto Lage, M.A.; Jimenez-Lopez, C.; Mejuto, J.C.; Simal-Gandara, J. The Potential of Seaweeds as a Source of Functional Ingredients of Prebiotic and Antioxidant Value. *Antioxidants* **2019**, *8*, 406. [CrossRef] [PubMed]
4. Cardozo, K.H.M.; Guaratini, T.; Barros, M.P.; Falcão, V.R.; Tonon, A.P.; Lopes, N.P.; Campos, S.; Torres, M.A.; Souza, A.O.; Colepicolo, P. Metabolites from Algae with Economical Impact. *Comp. Biochem. Physiol. Part C Toxicol. Pharmacol.* **2007**, *146*, 60–78. [CrossRef] [PubMed]
5. Dhanki, A.; Sindhav, S.; Jadeja, B.A. Evaluation of the Antimicrobial and Antioxidant Activity of Two Chlorophyceae and Two Rhodophyceae Seaweeds from Porbandar Coast. *Eur. J. Med. Plants* **2020**, *31*, 34–39. [CrossRef]
6. Tang, T.; Effiong, K.; Hu, J.; Li, C.; Xiao, X. Chemical Prevention and Control of the Green Tide and Fouling Organism Ulva: Key Chemicals, Mechanisms, and Applications. *Front. Mar. Sci.* **2021**, *8*, 618950. [CrossRef]
7. Bayomy, H.M. Effects of Culinary Treatments on the Physicochemical Properties of Ulva Lactuca Collected from Tabuk Coast of Red Sea in Saudi Arabia. *Saudi J. Biol. Sci.* **2022**, *29*, 2355–2362. [CrossRef]
8. Din, N.A.S.; Mohd Alayudin, A.S.; Sofian-Seng, N.S.; Rahman, H.A.; Mohd Razali, N.S.; Lim, S.J.; Wan Mustapha, W.A. Brown Algae as Functional Food Source of Fucoxanthin: A Review. *Foods* **2022**, *11*, 2235. [CrossRef]
9. Bringloe, T.T.; Starko, S.; Wade, R.M.; Vieira, C.; Kawai, H.; De Clerck, O.; Cock, J.M.; Coelho, S.M.; Destombe, C.; Valero, M. Phylogeny and Evolution of the Brown Algae. *CRC Crit. Rev. Plant Sci.* **2020**, *39*, 281–321. [CrossRef]
10. Rioux, L.-E.; Beaulieu, L.; Turgeon, S.L. Seaweeds: A Traditional Ingredients for New Gastronomic Sensation. *Food Hydrocoll.* **2017**, *68*, 255–265. [CrossRef]
11. Roy, S.; Anantharaman, P. Biochemical Compositions of Seaweeds Collected from Olaikuda and Vadakkadu, Rameshwaram, Southeast Coast of India. *J. Mar. Sci. Res. Dev.* **2017**, *7*, 1–5. [CrossRef]

12. Fitzgerald, C.; Gallagher, E.; Tasdemir, D.; Hayes, M. Heart Health Peptides from Macroalgae and Their Potential Use in Functional Foods. *J. Agric. Food Chem.* **2011**, *59*, 6829–6836. [CrossRef]
13. Taskin, O.S.; Ersoy, N.; Aksu, A.; Kiskan, B.; Balkis, N.; Yagci, Y. Melamine-Based Microporous Polymer for Highly Efficient Removal of Copper(II) from Aqueous Solution. *Polym. Int.* **2016**, *65*, 439–445. [CrossRef]
14. Osuna-Ruiz, I.; López-Saiz, C.-M.; Burgos-Hernández, A.; Velázquez, C.; Nieves-Soto, M.; Hurtado-Oliva, M.A. Antioxidant, Antimutagenic and Antiproliferative Activities in Selected Seaweed Species from Sinaloa, Mexico. *Pharm. Biol.* **2016**, *54*, 2196–2210. [CrossRef] [PubMed]
15. Aksu, A.; Balkis, N.; Taşkin, Ö.S.; Erşan, M.S. Toxic Metal (Pb, Cd, As and Hg) and Organochlorine Residue Levels in Hake (Merluccius Merluccius) from the Marmara Sea, Turkey. *Environ. Monit Assess* **2011**, *182*, 509–521. [CrossRef] [PubMed]
16. Toyosaki, T.; Iwabuchi, M. New Antioxidant Protein in Seaweed (Porphyra Yezoensis Ueda). *Int. J. Food Sci. Nutr.* **2009**, *60*, 46–56. [CrossRef] [PubMed]
17. Nagayama, K.; Iwamura, Y.; Shibata, T.; Hirayama, I.; Nakamura, T. Bactericidal Activity of Phlorotannins from the Brown Alga Ecklonia Kurome. *J. Antimicrob. Chemother.* **2002**, *50*, 889–893. [CrossRef] [PubMed]
18. Fernando, I.P.S.; Kim, M.; Son, K.-T.; Jeong, Y.; Jeon, Y.-J. Antioxidant Activity of Marine Algal Polyphenolic Compounds: A Mechanistic Approach. *J. Med. Food* **2016**, *19*, 615–628. [CrossRef] [PubMed]
19. Zubia, M.; Robledo, D.; Freile-Pelegrin, Y. Antioxidant Activities in Tropical Marine Macroalgae from the Yucatan Peninsula, Mexico. *J. Appl. Phycol.* **2007**, *19*, 449–458. [CrossRef]
20. Kohen, R.; Nyska, A. Invited Review: Oxidation of Biological Systems: Oxidative Stress Phenomena, Antioxidants, Redox Reactions, and Methods for Their Quantification. *Toxicol. Pathol.* **2002**, *30*, 620–650. [CrossRef] [PubMed]
21. Dawes, C.J. *Marine Botany*; John Wiley & Sons: Hoboken, NJ, USA, 1998; ISBN 0471192082.
22. Fleurence, J. Seaweed Proteins: Biochemical, Nutritional Aspects and Potential Uses. *Trends Food Sci. Technol.* **1999**, *10*, 25–28. [CrossRef]
23. Černá, M. Seaweed Proteins and Amino Acids as Nutraceuticals. *Adv. Food Nutr. Res.* **2011**, *64*, 297–312. [PubMed]
24. Ścieszka, S.; Klewicka, E. Algae in Food: A General Review. *Crit. Rev. Food Sci. Nutr.* **2019**, *59*, 3538–3547. [CrossRef] [PubMed]
25. Bhatnagar, P.; Gururani, P.; Parveen, A.; Gautam, P.; Chandra Joshi, N.; Tomar, M.S.; Nanda, M.; Vlaskin, M.S.; Kumar, V. Algae: A Promising and Sustainable Protein-Rich Food Ingredient for Bakery and Dairy Products. *Food Chem.* **2024**, *441*, 138322. [CrossRef] [PubMed]
26. Babich, O.; Sukhikh, S.; Larina, V.; Kalashnikova, O.; Kashirskikh, E.; Prosekov, A.; Noskova, S.; Ivanova, S.; Fendri, I.; Smaoui, S.; et al. Algae: Study of Edible and Biologically Active Fractions, Their Properties and Applications. *Plants* **2022**, *11*, 780. [CrossRef] [PubMed]
27. Pangestuti, R.; Kim, S.-K. Chapter 6-Seaweed Proteins, Peptides, and Amino Acids. In *Seaweed Sustainability*; Tiwari, B.K., Troy, D.J., Eds.; Academic Press: San Diego, CA, USA, 2015; pp. 125–140, ISBN 978-0-12-418697-2.
28. Yanshin, N.; Kushnareva, A.; Lemesheva, V.; Birkemeyer, C.; Tarakhovskaya, E. Chemical Composition and Potential Practical Application of 15 Red Algal Species from the White Sea Coast (the Arctic Ocean). *Molecules* **2021**, *26*, 2489. [CrossRef]
29. Norziah, M.H.; Ching, C.Y. Nutritional Composition of Edible Seaweed Gracilaria Changgi. *Food Chem.* **2000**, *68*, 69–76. [CrossRef]
30. Bhuiyan, K.A.; Qureshi, S.; Mustafa Kamal, A.H.; AftabUddin, S.; Siddique, A. Proximate Chemical Composition of Sea Grapes Caulerpa Racemosa (J. Agardh, 1873) Collected from a Sub-Tropical Coast. *Virol. Mycol.* **2016**, *5*, 2161–2517.
31. Lorenzo, J.M.; Agregán, R.; Munekata, P.E.S.; Franco, D.; Carballo, J.; Şahin, S.; Lacomba, R.; Barba, F.J. Proximate Composition and Nutritional Value of Three Macroalgae: Ascophyllum Nodosum, Fucus Vesiculosus and Bifurcaria Bifurcata. *Mar. Drugs* **2017**, *15*, 360. [CrossRef]
32. Denis, C.; Morançais, M.; Li, M.; Deniaud, E.; Gaudin, P.; Wielgosz-Collin, G.; Barnathan, G.; Jaouen, P.; Fleurence, J. Study of the Chemical Composition of Edible Red Macroalgae Grateloupia Turuturu from Brittany (France). *Food Chem.* **2010**, *119*, 913–917. [CrossRef]
33. Sánchez-Machado, D.I.; López-Cervantes, J.; López-Hernández, J.; Paseiro-Losada, P. Fatty Acids, Total Lipid, Protein and Ash Contents of Processed Edible Seaweeds. *Food Chem.* **2004**, *85*, 439–444. [CrossRef]
34. Biancarosa, I.; Belghit, I.; Bruckner, C.G.; Liland, N.S.; Waagbø, R.; Amlund, H.; Heesch, S.; Lock, E. Chemical Characterization of 21 Species of Marine Macroalgae Common in Norwegian Waters: Benefits of and Limitations to Their Potential Use in Food and Feed. *J. Sci. Food Agric.* **2018**, *98*, 2035–2042. [CrossRef] [PubMed]
35. Chan, P.T.; Matanjun, P. Chemical Composition and Physicochemical Properties of Tropical Red Seaweed, Gracilaria Changii. *Food Chem.* **2017**, *221*, 302–310. [CrossRef] [PubMed]
36. Kumar, M.; Kumari, P.; Trivedi, N.; Shukla, M.K.; Gupta, V.; Reddy, C.R.K.; Jha, B. Minerals, PUFAs and Antioxidant Properties of Some Tropical Seaweeds from Saurashtra Coast of India. *J. Appl. Phycol.* **2011**, *23*, 797–810. [CrossRef]
37. López-López, I.; Cofrades, S.; Ruiz-Capillas, C.; Jiménez-Colmenero, F. Design and Nutritional Properties of Potential Functional Frankfurters Based on Lipid Formulation, Added Seaweed and Low Salt Content. *Meat Sci.* **2009**, *83*, 255–262. [CrossRef] [PubMed]
38. Rupérez, P.; Ahrazem, O.; Leal, J.A. Potential Antioxidant Capacity of Sulfated Polysaccharides from the Edible Marine Brown Seaweed Fucus Vesiculosus. *J. Agric. Food Chem.* **2002**, *50*, 840–845. [CrossRef] [PubMed]
39. Murray, M.; Dordevic, A.L.; Ryan, L.; Bonham, M.P. An Emerging Trend in Functional Foods for the Prevention of Cardiovascular Disease and Diabetes: Marine Algal Polyphenols. *Crit. Rev. Food Sci. Nutr.* **2018**, *58*, 1342–1358. [CrossRef]

40. Chkhikvishvili, I.D.; Ramazanov, Z.M. Phenolic Substances of Brown Algae and Their Antioxidant Activity. *Appl. Biochem. Microbiol.* **2000**, *36*, 289–291. [CrossRef]
41. Kang, K.; Park, Y.; Hwang, H.J.; Kim, S.H.; Lee, J.G.; Shin, H.-C. Antioxidative Properties of Brown Algae Polyphenolics and Their Perspectives as Chemopreventive Agents against Vascular Risk Factors. *Arch. Pharm. Res.* **2003**, *26*, 286–293. [CrossRef]
42. Duan, X.-J.; Zhang, W.-W.; Li, X.-M.; Wang, B.-G. Evaluation of Antioxidant Property of Extract and Fractions Obtained from a Red Alga, Polysiphonia Urceolata. *Food Chem.* **2006**, *95*, 37–43. [CrossRef]
43. Cho, M.; Kang, I.-J.; Won, M.-H.; Lee, H.-S.; You, S. The Antioxidant Properties of Ethanol Extracts and Their Solvent-Partitioned Fractions from Various Green Seaweeds. *J. Med. Food* **2010**, *13*, 1232–1239. [CrossRef]
44. Shahidi, F.; Zhong, Y. 18-Antioxidants from Marine by-Products. In *Maximising the Value of Marine By-Products*; Shahidi, F., Ed.; Woodhead Publishing: Sawston, UK, 2007; pp. 397–412, ISBN 978-1-84569-013-7.
45. Siriwardhana, N.; Lee, K.-W.; Jeon, Y.-J.; Kim, S.-H.; Haw, J.-W. Antioxidant Activity of Hizikia Fusiformis on Reactive Oxygen Species Scavenging and Lipid Peroxidation Inhibition. *Food Sci. Technol. Int.* **2003**, *9*, 339–346. [CrossRef]
46. Milinovic, J.; Campos, B.; Mata, P.; Diniz, M.; Noronha, J.P. Umami Free Amino Acids in Edible Green, Red, and Brown Seaweeds from the Portuguese Seashore. *J. Appl. Phycol.* **2020**, *32*, 3331–3339. [CrossRef]
47. Mæhre, H.K.; Malde, M.K.; Eilertsen, K.-E.; Elvevoll, E.O. Characterization of Protein, Lipid and Mineral Contents in Common Norwegian Seaweeds and Evaluation of Their Potential as Food and Feed. *J. Sci. Food Agric.* **2014**, *94*, 3281–3290. [CrossRef]
48. Cofrades, S.; López-Lopez, I.; Bravo, L.; Ruiz-Capillas, C.; Bastida, S.; Larrea, M.T.; Jiménez-Colmenero, F. Nutritional and Antioxidant Properties of Different Brown and Red Spanish Edible Seaweeds. *Food Sci. Technol. Int.* **2010**, *16*, 361–370. [CrossRef] [PubMed]
49. Horwitz, W. *Official Methods of Analysis*; Association of Official Analytical Chemists: Washington, DC, USA, 1975; Volume 222.
50. Bharathi, S.; Dinesh Kumar, S.; Sekar, S.; Santhanam, P.; Divya, M.; Krishnaveni, N.; Pragnya, M.; Dhanalakshmi, B. Experimental Evaluation of Seaweeds Liquid Extracts as an Alternative Culture Medium on the Growth and Proximate Composition of Picochlorum Maculatum. *Proc. Natl. Acad. Sci. India Sect. B Biol. Sci.* **2021**, *91*, 205–215. [CrossRef]
51. Hemalatha, A.; Girija, K.; Parthiban, C.; Saranya, C.; Anantharaman, P. Antioxidant Properties and Total Phenolic Content of a Marine Diatom, Navicula Clavata and Green Microalgae, Chlorella Marina and Dunaliella Salina. *Adv. Appl. Sci. Res* **2013**, *4*, 151–157.
52. Dang, T.T.; Bowyer, M.C.; Van Altena, I.A.; Scarlett, C.J. Optimum Conditions of Microwave-Assisted Extraction for Phenolic Compounds and Antioxidant Capacity of the Brown Alga Sargassum Vestitum. *Sep. Sci. Technol.* **2018**, *53*, 1711–1723. [CrossRef]
53. Pulido, R.; Bravo, L.; Saura-Calixto, F. Antioxidant Activity of Dietary Polyphenols As Determined by a Modified Ferric Reducing/Antioxidant Power Assay. *J. Agric. Food Chem.* **2000**, *48*, 3396–3402. [CrossRef]

molecules

MDPI

Article

Polysaccharides from *Brasenia schreberi* with Great Antioxidant Ability and the Potential Application in Yogurt

Yujie Wang [1,2,†], Yue Zou [1,2,3,†], Qiong Fang [1,2], Ruizhang Feng [1,2], Jihong Zhang [1,2], Wanhai Zhou [1,2,4,*] and Qin Wei [1,2,*]

1 Faculty of Agriculture, Forestry and Food Engineering, Yibin University, Yibin 644000, China; wangyjsicau@126.com (Y.W.); zouyue255@163.com (Y.Z.); fangfang_0717@163.com (Q.F.); ruizhangfeng@126.com (R.F.); ysl199805@hotmail.com (J.Z.)

2 Sichuan Oil Cinnamon Engineering Technology Research Center, Yibin 644000, China

3 School of Food and Bioengineering, Xihua University, Chengdu 610039, China

4 Beijing Advanced Innovation Center for Food Nutrition and Human Health, Beijing Technology and Business University (BTBU), Beijing 100048, China

* Correspondence: wanhaizhou@126.com (W.Z.); weiqin2001-67@163.comail.com (Q.W.)

† These authors contributed equally to this work.

Abstract: *Brasenia schreberi* is a widely consumed aquatic plant, yet the knowledge regarding its bioactive components, particularly polysaccharides, remains limited. Therefore, this study aimed to optimize the extraction process of polysaccharides from *B. schreberi* using the response surface method (RSM). Additionally, we characterized the polysaccharides using various methods and assessed their antioxidant capabilities both in vitro and in vivo, employing cell cultures and *Caenorhabditis elegans*. Furthermore, these polysaccharides were incorporated into a unique yogurt formulation. Our findings demonstrated that hot water extraction was the most suitable method for extracting polysaccharides from *B. schreberi*, yielding samples with high sugar content, significant antioxidant capacity, and a well-defined spatial structure. Moreover, pectinase was employed for polysaccharide digestion, achieving an enzymolysis rate of 10.02% under optimized conditions using RSM. Notably, the results indicated that these polysaccharides could protect cells from oxidative stress by reducing apoptosis. Surprisingly, at a concentration of 250 μg/mL, the polysaccharides significantly increased the survival rate of *C. elegans* from 31.05% to 82.3%. Further qPCR results revealed that the polysaccharides protected *C. elegans* by up-regulating the *daf-16* gene and down-regulating mTOR and insulin pathways, demonstrating remarkable antioxidant abilities. Upon addition to the yogurt, the polysaccharides significantly enhanced the water retention, viscosity, and viability of lactic acid bacteria. These outcomes underscore the potential of polysaccharides from *B. schreberi* as a valuable addition to novel yogurt formulations, thereby providing additional theoretical support for the utilization of *B. schreberi*.

Keywords: *Brasenia schreberi*; enzymatic hydrolysis; polysaccharides; antioxidant ability; polysaccharide yogurt

1. Introduction

Brasenia schreberi, a perennial aquatic herb belonging to the Nymphaeaceae family, is currently under the threat of extinction, with its distribution confined to subtropical regions located south of 30° N, including countries such as China, Japan, South Korea, Australia, North America, Cuba, and Mexico [1]. In traditional folklore, *B. schreberi* was used as a dietary salad ingredient. Recent studies have highlighted its wide-ranging properties, demonstrating not only antioxidant and anti-inflammatory attributes but also revealing an inhibitory effect on HIV [1,2]. These findings have sparked significant scientific interest, particularly in the fields of nutritional and pharmaceutical research.

However, it is essential to note certain limitations in the research focused on polysaccharides derived from *B. schreberi*. While polysaccharides are widely recognized as the most abundant and main bioactive constituents in *B. schreberi* [3,4], current research has primarily highlighted their cholesterol-lowering effects and strong antioxidant abilities in vitro [5,6]. Despite these valuable discoveries, further investigations are required to comprehensively understand and harness the full spectrum of potential therapeutic applications associated with these polysaccharides. Future research efforts should aim to explore additional biological activities, elucidate underlying mechanisms, and investigate potential synergistic effects with other bioactive compounds present in *B. schreberi*. Such endeavors may provide a more comprehensive understanding of the diverse health-promoting properties of this botanical species, further enhancing its significance in both medical and nutritional contexts.

Numerous studies have demonstrated that plant polysaccharides have the potential to stimulate the growth of probiotics and regulate the enhancement of intestinal flora [7,8]. Particularly, polysaccharides derived from *Ganoderma lucidum* have been found to significantly increase the abundance of probiotic strains such as *Bifidobacterium*, *Lactobacillus* Johnson, and *Lactococcus lactis* in mice [9]. Likewise, polysaccharides isolated from *Poria cocos* have been shown to reduce the diversity of operational taxonomic units (OTUs) and profoundly remodel the composition of gut microbiota, favoring bacteria associated with anti-obesity effects, short-chain fatty acid production, and lactic acid production [9]. The recent literature has also indicated that the incorporation of polysaccharides during the fermentation process of yogurt can improve its texture and taste [10]. More specifically, polysaccharides can enhance the water-holding capacity of yogurt, fortify the protein-based network structure, and boost the stability of yogurt. Furthermore, the proliferation of lactic acid bacteria is facilitated [11]. Currently, some functional polysaccharides sourced from *Ganoderma lucidum*, *Tricholoma matsutake*, *Pleurotus eryngii*, and Citrus peel have been utilized as additives in yogurt production, yielding a distinctive and nutritious food product [12,13].

Degradation of polysaccharides into small molecular polysaccharides is an effective way to improve bioactivity. Likewise, polysaccharides from *Polygonatum sibiricum* and *Auricularia auricula* were degraded and then the degradation products surprisingly showed a better antioxidant ability [14,15]. Nowadays, due to the high-water solubility, the polysaccharides can be added to make functional fermented yogurt, which is also one of the development trends of the current dairy market [12,13]. At present, the research about polysaccharides from *B. schreberi* is more focused on the extraction method and structure composition [4,16]. Few researchers reported the degradation process of polysaccharides from *B. schreberi*, and the addition of degraded polysaccharides into food to make functional food, as most of their works are focused on the antioxidant ability in vitro [5].

Therefore, in this study, the aim was to explore the optimization process of enzymolysis of the polysaccharide from *B. schreberi*, evaluate the antioxidant ability of the polysaccharide, and to develop a high-quality yogurt with unique flavor and function using the obtained polysaccharide degradation products as an addition. This provided technical and theoretical support for the development of functional polysaccharide yogurt and enriched the research route of products from *B. schreberi*.

2. Results and Discussions

2.1. Chemical Properties of Polysaccharides Extracted from Different Conditions

As shown in Table 1, the total sugar content in water extraction polysaccharides (W-BSP) was the highest with 60.86% while alkaline extraction polysaccharides (A-BSP) had the least with 35.06%. The content of crude polysaccharides obtained using enzymatic extraction was in the middle with 54.45%, which was slightly lower than that in W-BSP, indicating that the pure enzymatic extraction of polysaccharides could not obtain higher yield, and the combination with other extraction methods might help to improve the yield. In addition, the reducing sugar content of polysaccharides extracted using the enzymatic method was the highest (enzymatic extraction polysaccharides, E-BSP, 1.44%). The three polysaccharides were all obtained after alcohol precipitation, and most reducing sugars in the extract were removed in 80% ethanol solution. Therefore, the reducing sugar content of polysaccharides obtained using the three extraction methods was low. It can be seen from Table 1, the protein content in A-BSP (9.45%) was much higher than that in E-BSP and W-BSP. The three extraction methods of polysaccharides yielded negative results in the iodine-potassium iodide reaction results, and the results of the ferric chloride reaction were also negative, meaning starch and phenolic hydroxyl did not exist in the three polysaccharides.

Table 1. Comparison of polysaccharide yield using different extraction methods.

Chemical Properties	Polysaccharides Extracted Using Different Methods		
	A-BSP	E-BSP	W-BSP
Total sugar content (%)	35.06 ± 0.12	54.45 ± 0.50	60.86 ± 0.41
Reducing sugar content (%)	0.22 ± 0.01	1.44 ± 0.02	0.36 ± 0.01
Protein content (%)	9.45 ± 0.18	1.56 ± 0.01	1.42 ± 0.03
Starch	-	-	-
Phenolic hydroxyl	-	-	-

Note: - means not detected.

The polysaccharides obtained using the three extraction methods were tasteless without phenolic hydroxyl and starch. Among them, the total sugar content of A-BSP (38.06%) was significantly lower than that of W-BSP (60.86%). In addition, the protein content (9.45%) was the highest and the color was the darkest. This may be beecause alkaline solution can dissolve not only a large number of neutral polysaccharides and acidic polysaccharides, but also substances that are insoluble in water. Moreover, the Maillard reaction can be promoted under alkaline conditions, resulting in the deepening of the color of the extract and difficult purification.

2.2. Triple Helix Structure of Different Polysaccharides

Congo red experiment can preliminarily determine whether there is a triple helix structure in polysaccharides [17]. As shown in Figure 1, when the concentration of sodium hydroxide increased gradually from 0 to 0.2 mol/L, the maximum absorption wavelength of W-BSP showed an obvious upward trend and reached the maximum at 0.2 mol/L, showing the red shift. After 0.2 mol/L, the maximum absorption wavelength decreased slowly, indicating that the triple helix structure was destroyed. Obviously, the maximum absorption wavelength of A-BSP was lower than that of the Congo red solution, meaning the triple helix structure did not exist in A-BSP. Although the red shift occurred in E-BSP, there was no downward trend at high concentrations of sodium hydroxide, hence it can be considered that there was no triple helix structure in E-BSP. Hence, the spatial structure of polysaccharides extracted by using hot water extraction was more complete and more suitable as the raw materials for degradation.

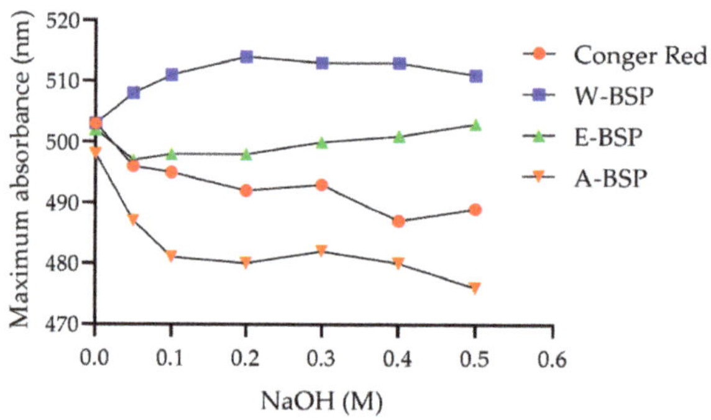

Figure 1. Change in the maximum absorption wavelength of polysaccharides.

2.3. The Antioxidant Capacity of Different Polysaccharides

As shown in Figure 2A, the DPPH· and ABTS free radical scavenging capacity was also gradually increased with the increasing concentration of crude polysaccharides. The IC50 values of A-BSP, E-BSP and W-BSP on DPPH were 0.55 mg/mL, 0.16 mg/mL and 0.23 mg/mL, respectively (Figure 2B). Meanwhile, The IC50 values of A-BSP, E-BSP and W-BSP on ABTS were 0.20 mg/mL, 0.15 mg/mL and 0.21 mg/mL, respectively. In addition, the absorbance of the three polysaccharides was increased, indicating an increased reducing power (Figure 2C). These results revealed that W-BSP possesses a strong antioxidant ability. Hence, in the following assays, W-BSP was selected for further enzymatic hydrolysis.

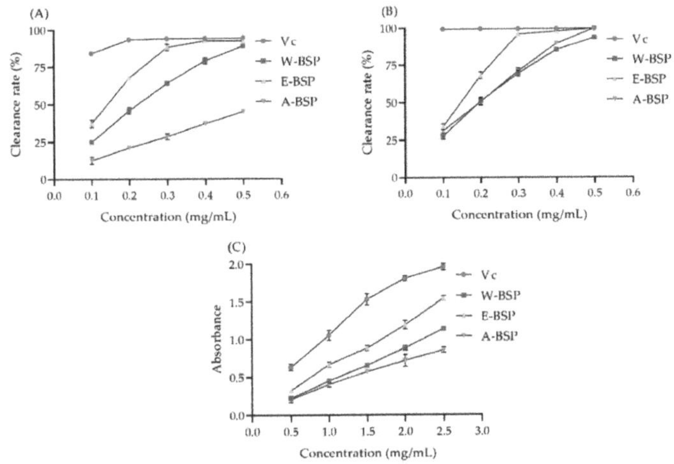

Figure 2. The antioxidant ability of the three polysaccharides. (**A**) The scavenging ability on (**A**) DPPH and (**B**) ABTS free radicals. And (**C**) the total reducing power.

2.4. Characteristics of Polysaccharides

After DEAE-52 column elution, the W-BSP-2 component accounted for 51.59%, which was the largest component (Figure 3A). After continuing to use the G-100 gel for elution, only one curve can be seen (Figure 3B), indicating that this part of the polysaccharide was a homogeneous polysaccharide. The infrared spectrum showed that the strong absorption peak at 3263.59 cm^{-1} was the result of the O-H stretching vibration of intermolecular hydrogen bonds [18]. The strong absorption peak at 2933 cm^{-1} was the C-H stretching vibration of methyl or methylene [19]. Near 1603 cm^{-1} was the characteristic peak of sugar hydrate [20]. 1414 cm^{-1} was the C-H changing angle vibration absorption peak of sugar (Figure 3C). After GPC-RI-MALLS measurement, the average relative molecular mass of the W-BSP-2 component was 6.19×10^5 (Figure 3D). The weight of the average relative molecular mass was 9.45×10^5. The root mean square radius of the rotation was 42.1 nm, and the weight root mean square (the radius of rotation) was 50 nm, and the polydispersity coefficient of W-BSP-2 was 1.525, indicating that the relative molecular weight distribution of the polysaccharide was narrow, and the molecular structure was relatively uniform. Following HPLC, it was determined that W-BSP-2 contained D-mannose, rhamnose, D-glucuronic acid, D-glucose, D-galactose, L-arabinose, and L-fucose 7 kinds of monosaccharides. The relative molar masses of the seven monosaccharides are as follows: 1; 1.86; 0.90; 0.06; 7.10; 1.39; and 3.82. And the percentages are 6.20%, 11.53%, 5.58%, 0.37%, 44.002%, 8.62%, and 23.68%, respectively. Under a 1000× microscope, the surface of W-BSP-2 was smooth and had a compact structure (Figure 3E). When the scanning electron microscope magnified it to 10,000×, it was observed that the surface of W-BSP-2 had a rough flaky structure with a corrugated structure, and the structure was still compact (Figure 3F).

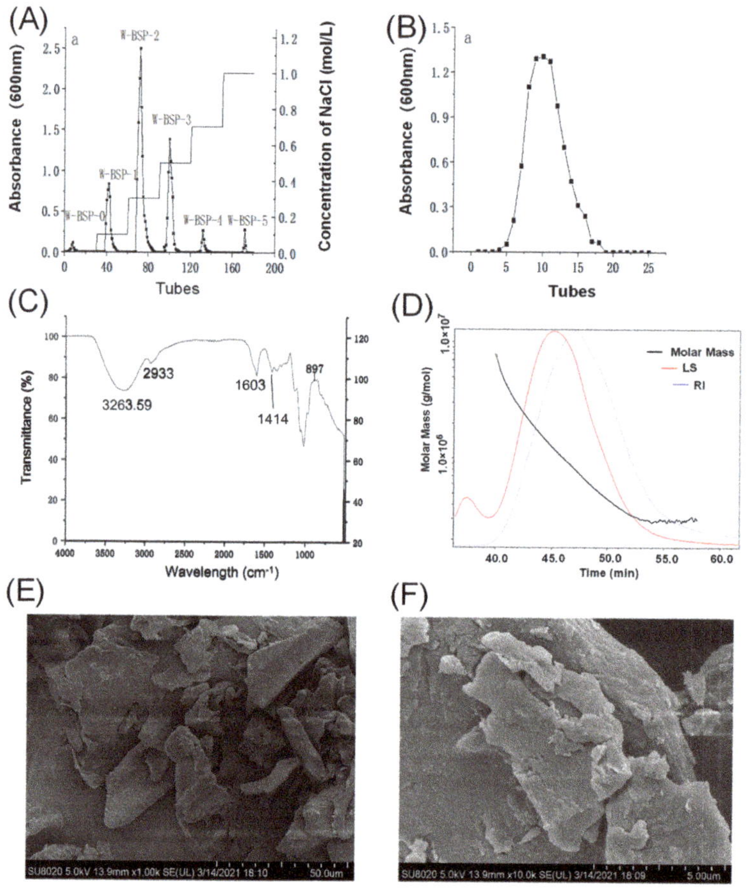

Figure 3. The characteristics of the polysaccharides. (**A**) Separation of the polysaccharides using DEAE Cellulose DE-52 gel; (**B**) separation and purification of Sephadex G-100 gel; (**C**) the Fourier Transform Infrared Spectroscopy spectra of the polysaccharides; (**D**) the GPC-RI-MALLS results of the polysaccharides; (**E**) SEM of the polysaccharides under 1000 magnification; and (**F**) SEM of the polysaccharides under 10,000 magnification.

2.5. Optimization of Enzymatic Hydrolysis

2.5.1. The Effect of Single Factors on Enzymatic Hydrolysis

Five kinds of enzymes were applied to the enzymatic hydrolyze polysaccharide under the optimum conditions, and the results showed that pectinase was the ideal degradation enzyme with the highest enzymatic hydrolysis rate of 4.65% (Supplementary Table S4). As shown in Figure 4A, the degradation rate was increased as the temperature increased, and the highest rate was 6.89% at 45 °C. A high temperature might destroy the structure of the enzymes. Hence 45 °C was selected as the optimum temperature for enzymatic hydrolysis. As the concentration of the enzyme increased to 60 U/mL, the degradation rate was promoted, and the maximum rate was 6.93% (Figure 4B). Continued increase in enzyme concentration did not increase the degradation rate of polysaccharides but lead to more enzyme residues. Therefore, 60 U/mL enzyme concentration was selected as the optimal enzyme concentration for enzymatic hydrolysis. Meanwhile, the optimal pH condition was easily found at 3.5 (Figure 4C). When the reaction time was 16 h, the enzymatic hydrolysis rate fluctuated around 10.02% (Figure 4D). Due to the use of the biological enzyme method for polysaccharide degradation, the conditions were very mild and the reaction speed was slow. Therefore, in the following RSM assay, the time was set at 2 h, and the time was set back at 16 h when the other conditions were optimized.

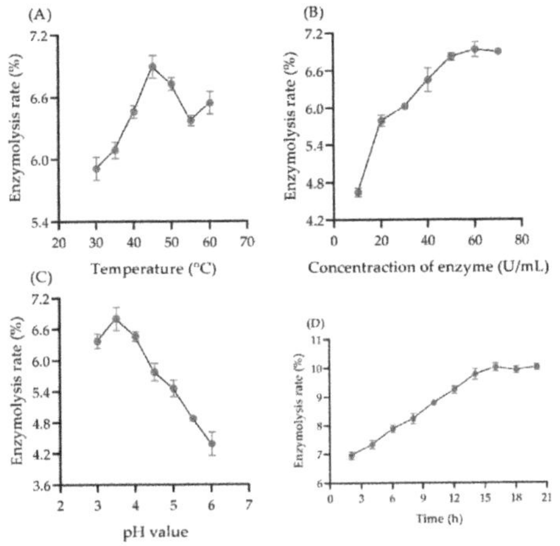

Figure 4. The effect of four single factors on the enzymolysis rate including (**A**) temperature, (**B**) concentration of enzyme, (**C**) pH value, and (**D**) fermentation time.

2.5.2. RSM Analysis

The equation for the enzymatic hydrolysis of polysaccharides was obtained through multivariate linear regression binomial fitting as follows:

$$Y = 6.799 + 0.519A + 0.004B + 0.055C + 0.034AB + 0.007AC - 0.001BC - 0.196A^2 - 0.354B^2 - 0.386C^2.$$

The R^2 was 0.9758, close to 1, meaning the model was significant. Variance results showed that the equation p value of the model was lower than 0.05. And the lack of fit was 0.2779 higher than 0.05 which was not significant (Table 2). The p value of A, A^2, B^2, and C^2 were lower than 0.05 which was extremely significant. As shown in Supplementary Figure S1, the interaction between enzyme concentration and temperature and pH value had great influence on the degradation rate of the polysaccharide, while the interaction between temperature and pH value had little influence on the degradation rate. According to the F value, the effect of various factors on the degradation of polysaccharides was enzyme concentration > temperature > pH value.

Table 2. Variance analysis of the regression equation.

Source	Square	Degree of Freedom	F Value	p Value
Model	3.65	9	31.39	<0.0001
A	2.16	1	167.12	<0.0001
B	1.21×10^{-4}	1	9.39×10^{-3}	0.9255
C	0.025	1	1.9	0.2104
AB	4.63×10^{-3}	1	0.36	0.5683
AC	1.93×10^{-4}	1	0.015	0.906
BC	6.90×10^{-6}	1	5.34×10^{-4}	0.9822
A^2	0.16	1	12.6	0.0093
B^2	0.53	1	40.96	0.0004
C^2	0.63	1	48.61	0.0002
Lack of fit	0.053	3	1.85	0.2779

Therefore, the model can be used to predict and analyze enzymatic hydrolysis. The optimal extraction conditions predicted were as follows: temperature was 45.80 °C, pH was 3.53, enzyme concentration was 60 U/mL, and the predicted degradation rate was 7.09%. In order to facilitate the experiment, the conditions were optimized as follows: temperature 45 °C, pH 3.5, and enzyme concentration 60 U/mL. Under these conditions, the average degradation rate of polysaccharides was 6.96%, which was not significantly different from the predicted value. Under the optimal conditions, the hydrolysis time was set back 16 h, and the average hydrolysis rate was 10.02%.

2.6. *The Antioxidant Ability of Polysaccharides In Vitro and In Vivo*

2.6.1. Polysaccharides Increased the Cell Viability under H_2O_2

As shown in Figure 5A, polysaccharides in a high concentration actually had a negative effect on the growth of HUVEC cells. Under 2000 μg/mL conditions, the cell viability was only 44.75% and the viability was 68.45% when the concentration was 1000 μg/mL. Therefore, for the following assays, we chose the lower toxic effect concentration to evaluate the protection of polysaccharides. H_2O_2 is an acknowledged trigger for oxidative stress at the cellular level [21,22]. More reactive oxygen species (ROS) will then be accumulated in the body, thus leading to side effects such as ageing, cancer, inflammatory, and so on [23–25]. In Figure 5B, the cell viability was only 64.06% when the cells were exposed to H_2O_2. When the cells were pre-treated with polysaccharides (125 and 250 μg/mL), the cell viability was significantly increased to 77.60% and 83.66%, respectively, while 500 μg/mL failed to enhance the cell viability ($p > 0.05$). Hence, these results demonstrated that polysaccharides in a low concentration showed low cytotoxicity and can protect cells against oxidative stress.

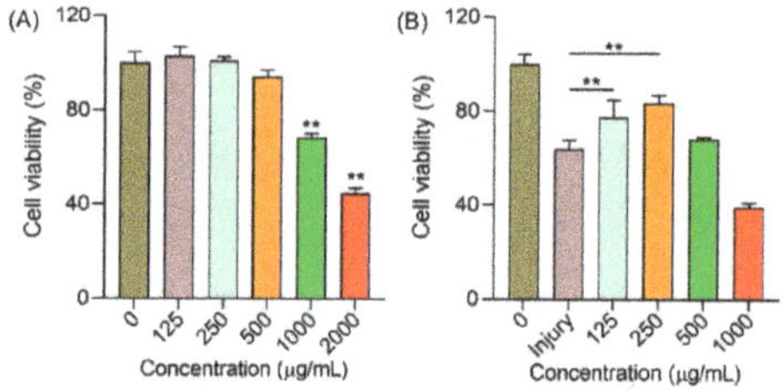

Figure 5. The effect of polysaccharides on HUVEC cells. (**A**) The cell viability of cells after treatment with polysaccharides for 24 h; and (**B**) the cell viability of cells that were exposed to H_2O_2. Note: ** means $p < 0.01$.

2.6.2. Polysaccharides Reduced Cell Apoptosis under H_2O_2

Under oxidative stress, cells will suffer cell apoptosis even death [26,27]. In this work, we used AO/EB and AO/PI, two staining methods to assess cell apoptosis. When the cells are alive, EB and PI do not go into the cells and so the PE signal channel is very low, while AO can stain alive cells making the FITC signal channel high. Therefore, the alive cells will be in the Q3 area (Figure 6A), the apoptosis cells will be in the Q4 area, and the dead cells will be in the Q1 area. In Figure 6B, treated with H_2O_2, the numbers of cells in Q3 and Q4 were increased. After treatment with polysaccharides, the apoptosis cells were reduced (Figure 6C,D). In the Q1 area, the number of cell deaths was reduced (Figure 6E) and the number of live cells increased after the cells were treated with polysaccharides (Figure 6F); while only 250 μg/mL concentration of polysaccharides could lower cell apoptosis (Figure 6G). Like other natural plants, the bioactive components can reduce cell apoptosis when the cells are exposed to oxidative stress [28,29].

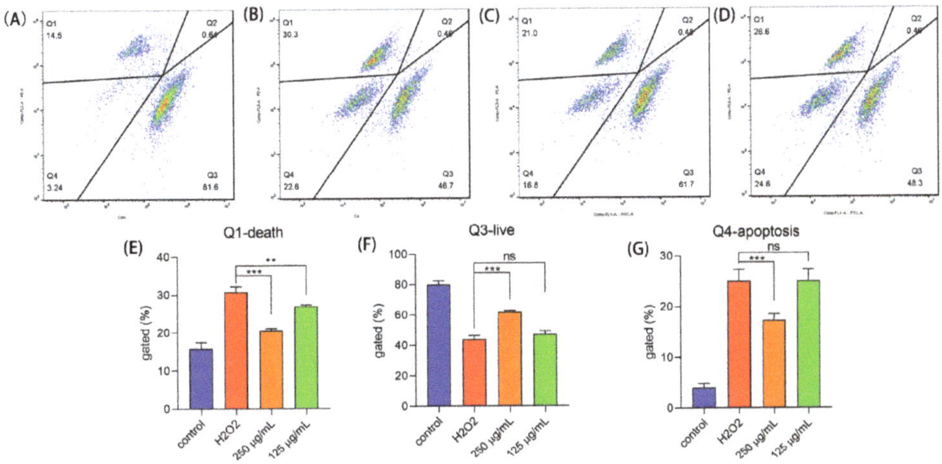

Figure 6. Cell apoptosis determined by AOEB. (**A**) The control group; (**B**) the group that was exposed to H_2O_2; (**C**) the group that was treated with 250 μg/mL; (**D**) the group that was treated with 125 μg/mL; and (**E–G**) the cell numbers in the Q1, Q3 and Q4 areas, respectively. Notes: *** means $p < 0.001$, ** means $p < 0.01$, and ns means not significant.

2.6.3. Polysaccharides Reduced ROS under H_2O_2

Then, we also detected the ROS level in the cells. Obviously, H_2O_2 could induce a high ROS level (Figure 7). Surprisingly, polysaccharides can lower the ROS level (Figure 6). Other previous works had already proved that H_2O_2 caused a high ROS level in cells which then induced a lower cell viability and cell apoptosis resulting in cell death [30]. In this work, we found that polysaccharide treatment can reduce the high ROS level and cell apoptosis. Hence, we drew one conclusion that polysaccharides increased the cell viability against oxidative stress through reducing the ROS level and cell apoptosis, showing a good antioxidant ability in vitro.

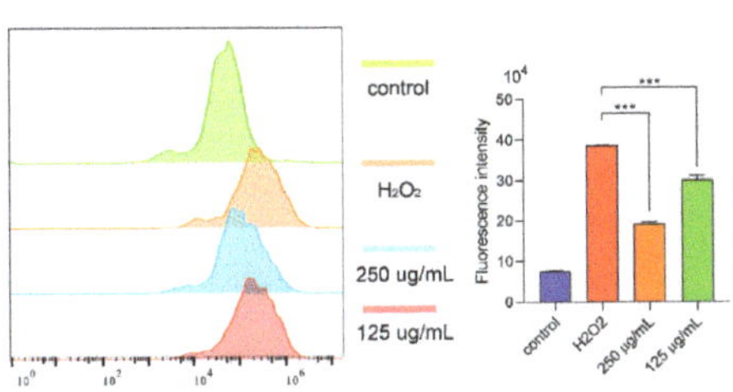

Figure 7. The ROS level in cells detected using flow cytometry. Note: *** means $p < 0.001$.

2.6.4. Polysaccharides Increased *C. elegans*' Survival under Thermal Stress

The *C. elegans* is a good research model to evaluate the pharmacological activity of natural products like polysaccharides, polyphenols, flavone, and so on [30–32]. We found polysaccharides showed excellent protection of cells against H_2O_2-induced oxidative stress. Hence, the *C. elegans* was applied to further assess the antioxidant ability of polysaccharides in vivo. As shown in Figure 8, the survival rate of the control group was 31.05%. 1000 μg/mL of polysaccharide treatment failed to extend the survival rate ($p > 0.05$), while the other three concentrations could all prolong the survival time of *C. elegans* under thermal stress. Among these, the effect of 250 μg/mL was the best, and the survival rate was increased to 82.3%, which was almost double than that in the control group, while when the concentration of polysaccharides was over 500 μg/mL, especially 1000 μg/mL, the survival of worms was not significantly changed. A high concentration of polysaccharides was also shown to have some toxic effects on HUVEC cells, hence a higher concentration may not be able to increase the survival of *C. elegans* due to the toxicity effect. Others natural polysaccharides also can protect *C. elegans* under thermal stress. Polysaccharides from the shells of *Camellia oleifera* increased the survival rate by 22.69% [30], and Fermented Coix Seed polysaccharides also enhanced the survival by 38.93% [33]. But, compared with the results in this work, polysaccharides exhibited an excellent protection of *C. elegans* under hot conditions.

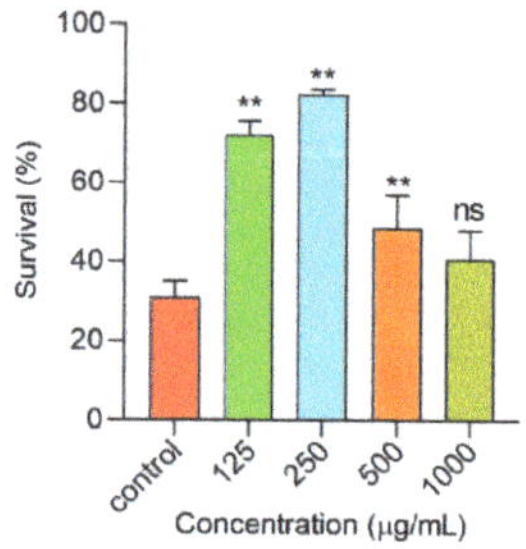

Figure 8. The worm survival under thermal stress. Notes: ** means $p < 0.01$, and ns means not significant.

2.6.5. Polysaccharides Reduced the ROS Level under Thermal Stress

Thermal stress can cause severe oxidative damage to *C. elegans* even leading to death [32]. Under hot conditions, the ROS level in *C. elegans* was significantly increased. After being treated with polysaccharides, the ROS level was then reduced (Figure 9). Hence, we can say that the polysaccharides can lower the ROS level to relieve the oxidative damage induced by thermal stress.

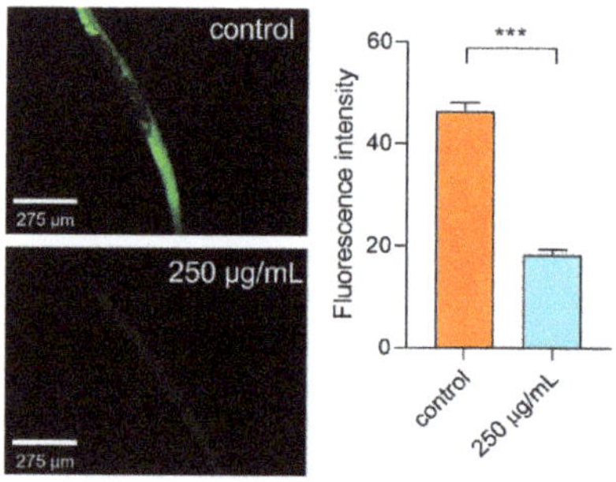

Figure 9. The ROS level in worms under thermal stress. Note: *** means $p < 0.001$.

2.6.6. Polysaccharides Regulated Gene Expression in the IIS/MAPK/mTOR Signal Pathway

Subsequently, we further assessed the expression of some genes that are involved in the defense to stress. For instance, the expression of DAF-16 was significantly increased (Figure 10A). DAF-16 is a key gene that regulates worms to defend stress [34]. While the expression of DAF-16 is regulated by other pathways like the insulin pathway (IIS) [35], surprisingly, the expression of DAF-2 was reduced (Figure 10B), meaning the polysaccharides can down-regulate the IIS pathway. Heat shock factor 1 (HSF-1) is a key transcription factor regulating heat stress and protein folding homeostasis. Therefore, we examined the expression of HSF-1. As shown in Figure 10C, the expression of HSF-1 was obviously increased. Hence the polysaccharides may activate HSF-1 to help worms against stress conditions. SKN-1 is an essential regulator of antioxidant activity and xenobiotic defense [36].

Glochidion zeylanicum leaf extracts and *Laminaria japonica* polysaccharide can activate SKN-1 in *C. elegans* under thermal stress to protect the worms [37,38]. In this work, we also found that the expression of SKN-1 was increased (Figure 10D). Moreover, the expressions of let-363 and clk-1 were suppressed (Figure 10E,F), demonstrating that the mTOR signal pathway was involved. The expression of aak-2 was also increased (Figure 10G), meaning the polysaccharides may up-regulate the AMPK signal pathway. PMK-1 is another important gene involved in the core p38 MAPK signaling pathway, and SKN-1 is the downstream target for PMK-1 when the worms respond to stress [39]. As shown in Figure 10H, the expression of pmk-1 was obviously increased. In all, polysaccharides can up-regulate the AMPK and MAPK pathway and down-regulate the IIS/mTOR pathways, and then finally activate DAF-16 to enhance the tolerance ability of *C. elegans*.

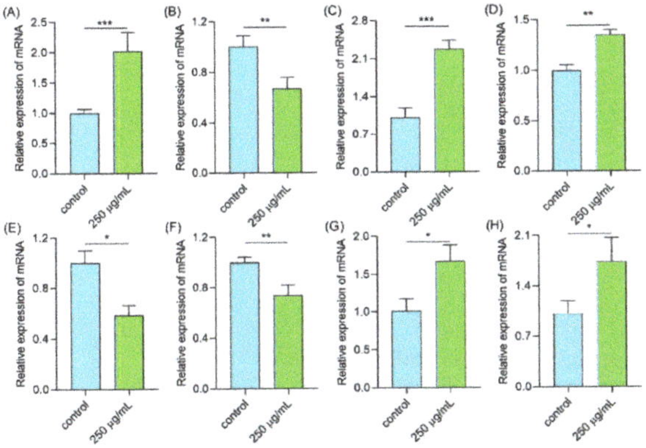

Figure 10. The expressions of genes in worms under thermal stress including (**A**) DAF-16, (**B**) DAF-2, (**C**) HSF-1, (**D**) SKN-1, (**E**) let-363, (**F**) clk-1, (**G**) aak-2, and (**H**) PMK-1. Notes: * means $p < 0.05$, ** means $p < 0.01$, and *** means $p < 0.001$.

2.6.7. Polysaccharides Up-Regulated Gene Expression Involved in the Antioxidant System

Under thermal stress, the ROS level in *C. elegans* was increased. In the organism, the antioxidant system is the main defense system to eliminate the high level of ROS. The sod-3 gene codes for the antioxidant enzyme, sod, in *C. elegans* [40]. Gst-4 and ctl-1 regulate the expression of GSH and CAT enzymes in *C. elegans*, respectively [41]. The above three enzymes are well-known as vital tools to eliminate the exceeding -level of ROS in the organism [42]. Therefore, we further determined the gene expression in the antioxidant system. As shown in Figure 11, the expressions of sod-3, gst-4, and ctl-1 were significantly increased. The three genes were downstream of DAF-16. As the expression of DAF-16 was promoted, the genes downstream of DAF-16 were then activated such as sod-3, gst-4, and ctl1. Hence, we can assume that the polysaccharides activate DAF-16, resulting in a high level of antioxidant enzymes which finally reduces the high ROS.

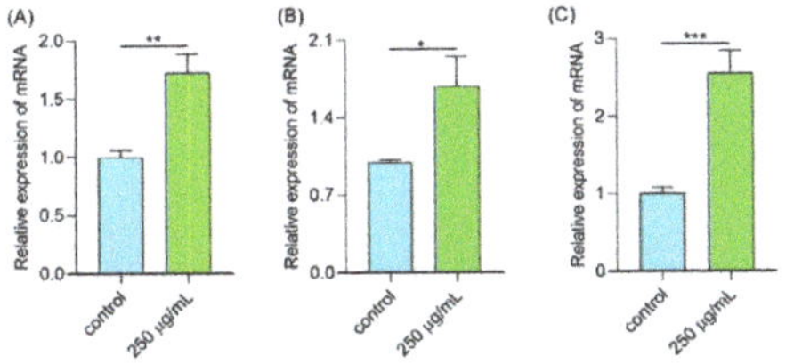

Figure 11. The expressions of antioxidant genes including (**A**) sod-3, (**B**) gst-4, and (**C**) ctl-1. Notes: * means $p < 0.05$, ** means $p < 0.01$, and *** means $p < 0.001$.

2.7. The Effect of Polysaccharides on Yogurt

2.7.1. Acidity and pH Value of Yogurt

The polysaccharide possesses a high antioxidant ability. Hence, we want to add this polysaccharide to make a unique yogurt. Acidity is one of the important physical and chemical indicators that need to be monitored in the production and storage of yogurt. pH value can reflect the acidity of yogurt under certain conditions [43]. As shown in Figure 12A, with the prolonged fermentation time, the titratable acid of fermented milk gradually increased, and it exceeded 70° T after 5 h. When the polysaccharide addition was 0.1%, the titratable acid was boosted, and the pH value decreased (Figure 12B). Therefore, fermentation and acid production was promoted when the polysaccharide concentration was low, while the fermentation of yogurt was inhibited when the added concentration was high.

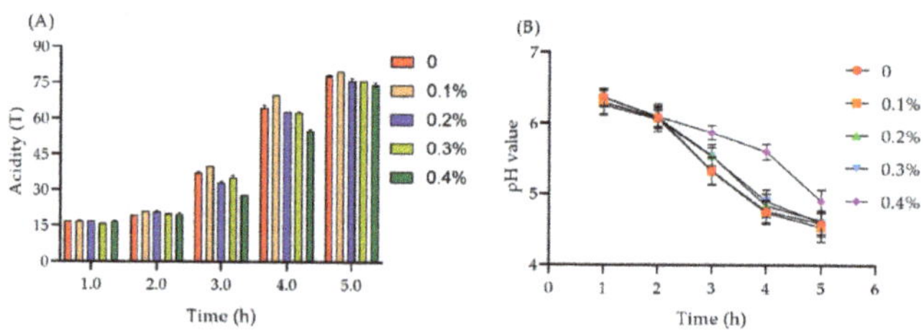

Figure 12. The effect of the addition of polysaccharides on the acidity and pH of yogurt during the process of fermentation. (**A**) the acidity of yogurt after adding polysaccharides for different time. (**B**) the pH changes of yogurt after adding polysaccharides for different time.

2.7.2. Water Holding Capacity of Yogurt

The water holding capacity of yogurt is mainly related to its gel structure. The stronger the gel structure is, the higher the water holding capacity is [44]. As shown in Figure 13, the water holding capacity of yogurt enhanced with the increase of polysaccharide concentration. The maximum capacity reached 92.19% when the polysaccharide addition was 0.2%, indicating that a low concentration of polysaccharide can promote the gel structure of yogurt and improve the water holding capacity of yogurt. When the addition of polysaccharides exceeded 0.2%, it was observed that the gel structure of yogurt was inhibited, resulting in a decrease in the water holding capacity of the yogurt.

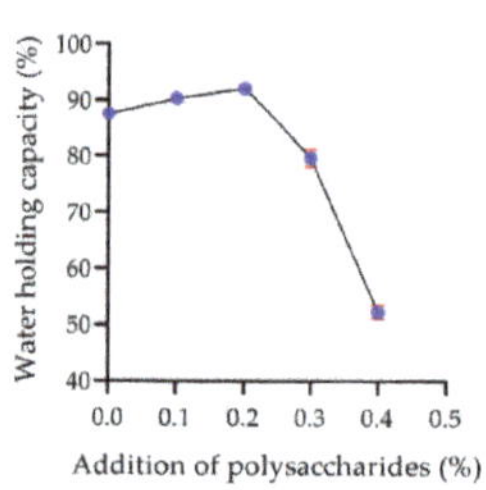

Figure 13. The effect of the addition of polysaccharides on the water holding capacity of yogurt.

2.7.3. Viscosity of Yogurt

As the polysaccharide concentration increased, the viscosity of the yogurt was promoted. When the polysaccharide concentration was 0.2%, the viscosity reached the maximum value of 234.67 mPa·s (Figure 14). However, when the addition of polysaccharides was more than 0.2%, the viscosity of yogurt decreased. When the addition of polysaccharides exceeds 0.2%, it may disrupt the gel structure of the yogurt, consequently reducing its viscosity.

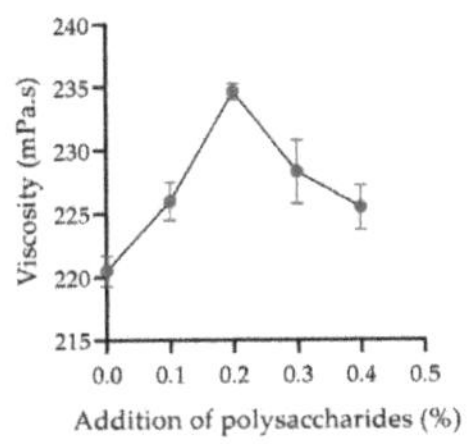

Figure 14. The effect of the addition of polysaccharides on viscosity.

2.7.4. Texture Property of Yogurt

With the increase of polysaccharide content, the elasticity and adhesion of yogurt increased at first but then decreased (Table 3). Elasticity makes the yogurt more resistant to chewing, and the adhesion makes the yogurt thick and rich in taste [44]. When the polysaccharide content was 0.2%, the elastic measurement value of the yogurt reached the maximum value of 8.94 mm, and when the polysaccharide content was 0.1%, the adhesion measurement value of the yogurt reached the maximum value of 3.02 mj, indicating that the appropriate amount of polysaccharide will increase the elasticity and adhesion of the yogurt. The hardness, adhesiveness and adhesion elongation of the yogurt decreased as more polysaccharide content was added, suggesting that over addition of polysaccharides could weaken the gel structure strength of the yogurt composed of casein.

Table 3. Effect of polysaccharide addition on yogurt texture characteristics.

Addition of Polysaccharides (%)	Hardness (N)	Elasticity (mm)	Glueyness (N)	Adhesivity (mj)	Adhesion Elongation (mm)
0	1.19 ± 0.005 [a]	6.93 ± 0.38 [a]	0.51 ± 0.012 [a]	2.68 ± 0.01 [a]	10.0 ± 0.089 [a]
0.1	1.15 ± 0.015 [b]	8.79 ± 0.08 [b]	0.50 ± 0.007 [a]	3.02 ± 0.04 [b]	5.90 ± 0.030 [b]
0.2	1.11 ± 0.006 [c]	8.94 ± 0.07 [b]	0.49 ± 0.003 [b]	2.82 ± 0.025 [c]	4.17 ± 0.032 [c]
0.3	0.98 ± 0.005 [d]	7.27 ± 0.16 [c]	0.44 ± 0.006 [c]	2.30 ± 0.032 [d]	3.31 ± 0.036 [d]
0.4	0.89 ± 0.01 [e]	5.87 ± 0.10 [d]	0.41 ± 0.002 [d]	2.18 ± 0.017 [e]	2.78 ± 0.043 [e]

Note: different letters mean different. In the same column, the different letters mean the significance like the value labeled with a is significantly different from that labeled with b.

2.7.5. Viable Lactic Acid Bacteria of Yogurt

Lactobacillus bulgaricus and *Streptococcus thermophilus* are the two most commonly used probiotics in yogurt production [45,46]. Probiotics, as a kind of living microorganisms that can give the host health benefits when a sufficient amount is applied, can improve human health by regulating disordered intestinal flora [47]. As shown in Table 4, the number of viable lactic acid bacteria in the polysaccharide yogurt increased significantly compared with the control group. When the polysaccharide content was 0.2%, the number of viable lactic acid bacteria reached the maximum of 7.50×10^8 CFU/mL. And in the above results, we found a higher concentration of polysaccharides showed some toxicity to cells and *C. elegans*, hence, if more polysaccharides were added in, the viability of bacteria may have been affected. Therefore, 0.2% addition is the ideal concentration.

Table 4. Effect of polysaccharide addition on the number of lactic acid bacteria.

Addition of Polysaccharides (%)	Lactic Acid Bacteria (CFU/mL)
0	$4.63 \times 10^8 \pm 0.35$ [a]
0.1	$6.82 \times 10^8 \pm 0.21$ [b]
0.2	$7.50 \times 10^8 \pm 0.10$ [c]
0.3	$6.20 \times 10^8 \pm 0.17$ [d]
0.4	$5.37 \times 10^8 \pm 0.36$ [e]

Note: different letters mean significant. In the same column, the different letters mean the significance like the value labeled with a is significantly different from that labeled with b.

2.7.6. Sensory Evaluation of Yogurt

As shown in Table 5, the sensory score increased as the polysaccharides were added, and the sensory score reached the highest when the added amount was 0.2%, indicating that adding a certain amount of polysaccharide can improve the texture and taste of yogurt. When the added amount was more than 0.2%, the sensory score significantly decreased, meaning that excessive polysaccharides can inhibit the fermentation of yogurt, and affect the coagulation, color and taste of the yogurt. Excessive polysaccharides may react with other components in the milk and inhibit the formation of the three-dimensional gel network structure between casein.

Table 5. Effect of polysaccharide addition on the sensory quality of yogurt.

Addition of Polysaccharides (%)	Sensory Evaluation	Scores
0	Uniform coagulation, moderate sweet and sour, milky white color, good taste, smell coordination	78.6 ± 1.18 [a]
0.1	Uniform coagulation, moderate sweet and sour, milky white color, thick taste, smell coordination	80.67 ± 1.64 [a]
0.2	Uniform coagulation, moderate sweet and sour, structure even, light yellow color, delicate and smooth taste, harmonious odor	81.11 ± 0.90 [a]
0.3	Coagulation milk, moderate acidity and sweetness, slightly yellow color, good taste, harmonious odor	65.89 ± 1.10 [b]
0.4	No coagulation, sweet and sour, slightly yellow in color, poor taste	43.89 ± 3.54 [c]

Note: different letters mean significant. In the same column, the different letters mean the significance like the value labeled with a is significantly different from that labeled with b.

2.8. The Effect of Single Factors on Yogurt Quality

2.8.1. Fermentation Temperature

Temperature affects the growth and metabolism of lactic acid bacteria, hence showing a significant impact on the quality of yogurt, mainly on the organizational structure, the flavor and taste of the yogurt, and a little effect on the color [48]. As shown in Figure 15, at a low temperature, the ability of lactic acid bacteria to produce acid decreased, resulting in the failure of coagulation of the yogurt fermented at 36 °C. At 38 °C, the taste of the yogurt was faint, and the flavor of the yogurt was not prominent, resulting in a low score. The yogurt fermented at 40–42 °C coagulated well, the acidity and viscosity of yogurt increased, and the sensory score increased. The sensory score at 42 °C was the highest. However, when the temperature was over 42 °C, the yogurt fermentation was excessive, leading to too much acidity, hence the sensory score was decreased.

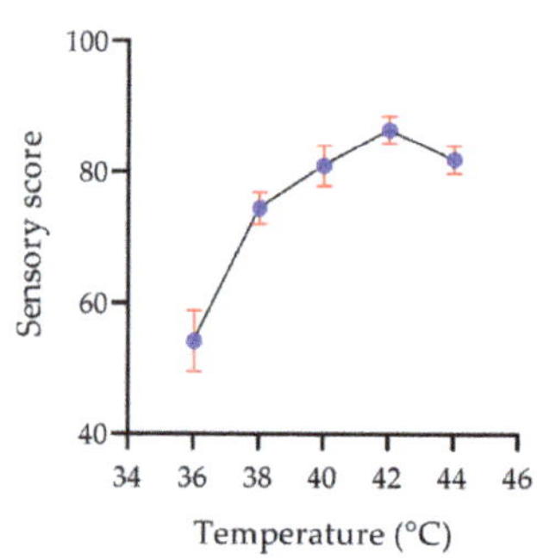

Figure 15. The effect of temperature on the sensory score of the yogurt.

2.8.2. Inoculation Amount

As shown in Figure 16, compared with other factors, the increase of the inoculation amount of bacterial powder exhibited little effect on the sensory score of the yogurt. When the inoculation amount was 0.10~0.12%, the quality of the yogurt did not change, and the highest sensory score was 84.4 at 0.12%. When the inoculation amount was higher than 0.12%, the sensory score of the yogurt decreased slowly as the inoculation amount increased.

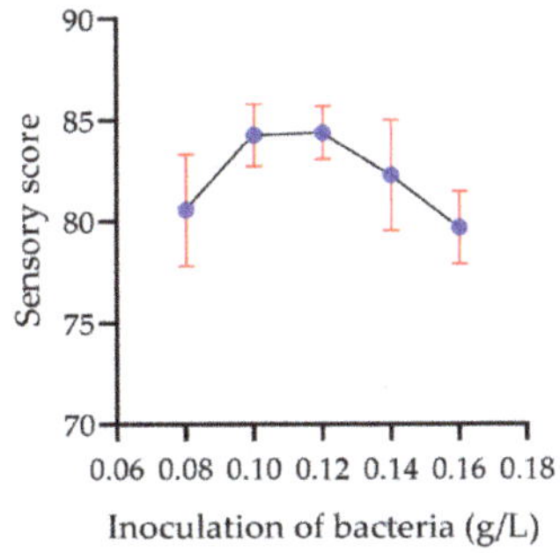

Figure 16. The effect of the inoculation of bacteria on the sensory score of the yogurt.

2.8.3. Polysaccharide Concentration

As the polysaccharide content was increased, the color of the yogurt changed from milk white to slightly yellow, and the taste was gradually viscous. When the polysaccharide concentration was 0.20%, the yogurt began to appear slightly astringent. Figure 17 shows that when the addition of polysaccharide was 0.15%, the color of the yogurt was normal, the taste was delicate, and there was no astringency with a maximum sensory score of 84.6.

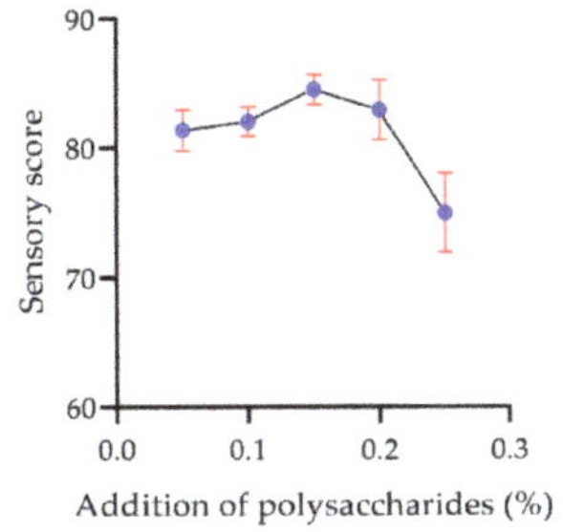

Figure 17. The effect of the addition of polysaccharides on the sensory score of the yogurt.

2.8.4. Fermentation Time

As shown in Figure 18, the sensory score of the yogurt increased with prolongation of fermentation time. When the fermentation time was short, the yogurt tasted faint and the curd was uneven. However, the yogurt was acidic, and the organization was slightly rough after a longer fermentation time. When the fermentation time was 5.5 h, the yogurt coagulated uniformly and had a delicate taste, and the sensory score of the yogurt was the highest at 86.6.

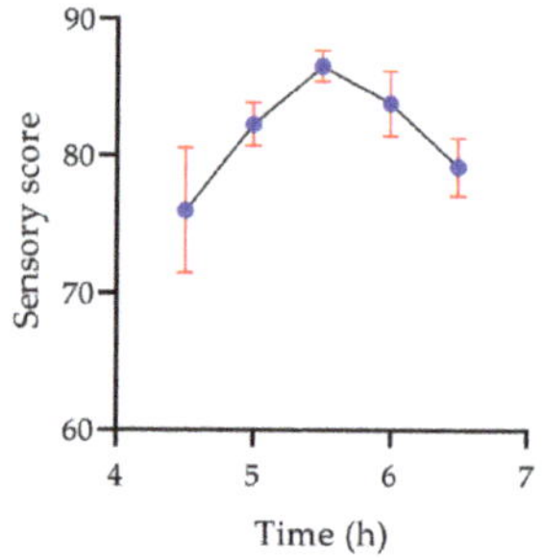

Figure 18. The effect of time on the sensory score of the yogurt.

2.8.5. Sucrose Concentration

In the fermentation of yogurt, the addition of sucrose is mainly to adjust its sweetness, so the addition of sucrose has little effect on the structure, color and flavor of the yogurt [49]. As shown in Figure 19, the sensory score of the yogurt increased as the sucrose content was added. When the sucrose content was low, the yogurt was slightly sour, while the yogurt was too sweet when the sucrose content was high. The sensory score of the yogurt reached the highest value of 85.4 with moderate sweetness when the sucrose content was 6%.

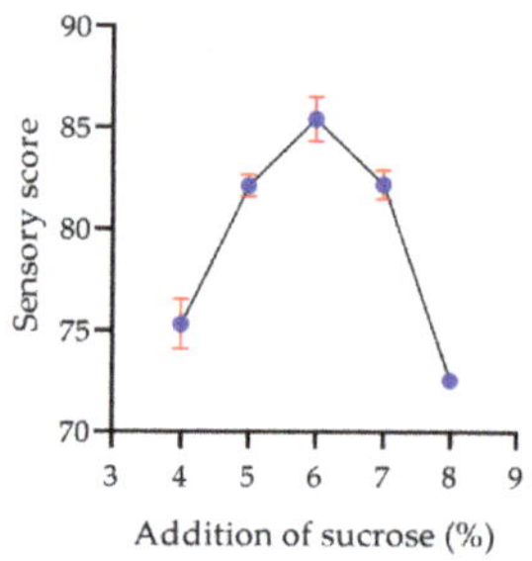

Figure 19. The effect of the addition of sucrose on the sensory score of the yogurt.

2.9. Orthogonal Optimization of Fermentation

Based on the results of single factor assays, three main factors were picked up for the orthogonal test. As shown in Supplementary Table S5, the factors affecting the fermentation of the polysaccharide yogurt were B (sucrose addition) > A (polysaccharide addition) > C (fermentation time). The optimal fermentation combination was selected as A2B2C2: 0.15% polysaccharide, 6% sucrose, and 5.5 h fermentation time.

2.10. Quality of Polysaccharide Yogurt after Orthogonal Optimization

Mold and yeast are the two most common types of bacterial contamination found in yogurt. *Staphylococcus aureus* and *Salmonella* are among the most typical food-borne pathogens. These bacteria not only affect the quality of yogurt but can also lead to food poisoning [50]. Therefore, these two pathogens are included as mandatory test items in the national standard for fermented milk. As demonstrated in Table 6, the viable count of lactic acid bacteria in polysaccharide yogurt exceeded 1 × 106 CFU/g, surpassing the threshold outlined in the national standard (China, GB 4789.1-2016) [51]. The counts for molds and yeasts fell within the national standard. Additionally, three other pathogens—*Staphylococcus aureus*, *Salmonella*, and *Escherichia coli*—were not detected in the yogurt samples. Illustrated in Table 7, the water holding capacity, acidity, and protein content of the yogurt were higher than those of regular yogurt, indicating that polysaccharide yogurt has the potential to enhance the nutritional value of yogurt.

Table 6. Bacterium standards of polysaccharide yogurt.

Strains	Standards	Determination
Lactic acid bacteria	$\geq 1 \times 10^6$ CFU/g	6.9×10^8
Escherichia coli	n = 5, c = 2, m = 1 CFU/g, M = 5 CFU/g	0
Staphylococcus aureus	0/25 g	0
Salmonella	0/25 g	0
Mold	$\leq$30 CFU/g	8
Yeast	$\leq$100 CFU/g	35

Table 7. Physical and chemical standards of polysaccharide yogurt.

Properties	Polysaccharides Yogurt	Ordinary Yogurt
Protein (g/100 g)	2.82 ± 0.04 ^{ns}	2.61 ± 0.03
Fat (g/100 g)	3.35 ± 0.02 ^{ns}	3.32 ± 0.02
Water holding capacity	$91.67 \pm 0.24\%$ *	$88.90 \pm 0.4\%$
Acidity (°T)	80.15 ± 2.17 *	74.32 ± 1.21
Sensory score	85.23 ± 2.37 *	81.48 ± 1.04

Note: ns means not significant and * means $p < 0.05$.

2.11. Simulated Digestion of Polysaccharide Yogurt In Vitro

The growth of lactic acid bacteria in yogurt in simulated intestinal fluid in vitro is presented in Table 8. The viable count of lactic acid bacteria in both regular yogurt and polysaccharide yogurt exhibited a declining trend at a pH of 1.5, indicating that excessive acidity in artificial gastric juice can impede the growth and reproduction of lactic acid bacteria. At pH levels of 2.5 and 3.5, the viable count of lactic acid bacteria demonstrated a pattern of decrease followed by an increase, suggesting that lactic acid bacteria can gradually adapt to artificial gastric juice at pH 3. Notably, at the same pH level, the number of lactic acid bacteria in polysaccharide yogurt exceeded that in regular yogurt, signifying superior acid resistance of lactic acid bacteria in polysaccharide yogurt compared to regular yogurt. As depicted in Table 9, within artificial simulated intestinal fluid, the viable lactic acid bacteria count in regular yogurt displayed a gradual decline, while the bacterial count in polysaccharide yogurt showed no significant change, indicating a better tolerance of lactic acid bacteria in polysaccharide yogurt to artificially simulated intestinal fluid.

Table 8. Growth of *lactobacillus* in yogurt in vitro gastric digestion.

pH Value	Time (h)	Lactic Acid Bacteria (CFU/mL)	
		Ordinary Yogurt	Polysaccharide Yogurt
1.5	0	$3.23 \times 10^8 \pm 0.22$	$3.61 \times 10^8 \pm 0.09$
	1	$3.01 \times 10^8 \pm 0.30$	$3.44 \times 10^8 \pm 0.14$
	2	$1.38 \times 10^8 \pm 0.15$	$2.42 \times 10^8 \pm 0.12$
	3	$1.19 \times 10^8 \pm 0.21$	$2.09 \times 10^8 \pm 0.45$
2.5	0	$4.17 \times 10^8 \pm 0.33$	$5.11 \times 10^8 \pm 0.27$
	1	$3.64 \times 10^8 \pm 0.54$	$3.82 \times 10^8 \pm 0.15$
	2	$4.02 \times 10^8 \pm 0.19$	$4.51 \times 10^8 \pm 0.39$
	3	$5.09 \times 10^8 \pm 0.60$	$5.64 \times 10^8 \pm 0.25$
3.5	0	$4.24 \times 10^8 \pm 0.22$	$5.00 \times 10^8 \pm 0.65$
	1	$3.67 \times 10^8 \pm 0.53$	$4.39 \times 10^8 \pm 0.38$
	2	$4.52 \times 10^8 \pm 0.27$	$4.63 \times 10^8 \pm 0.13$
	3	$4.81 \times 10^8 \pm 0.16$	$5.11 \times 10^8 \pm 0.26$

Table 9. Growth of lactobacillus in yogurt in vitro intestinal digestion.

Time (h)	Lactic Acid Bacteria (CFU/mL)	
	Ordinary Yogurt	**Polysaccharide Yogurt**
0	$6.11 \times 10^8 \pm 0.37$	$6.53 \times 10^8 \pm 0.25$
1	$5.93 \times 10^8 \pm 0.54$	$6.32 \times 10^8 \pm 0.18$
2	$5.62 \times 10^8 \pm 0.41$	$6.38 \times 10^8 \pm 0.28$
3	$5.69 \times 10^8 \pm 0.12$	$6.41 \times 10^8 \pm 0.45$
4	$5.31 \times 10^8 \pm 0.14$	$6.51 \times 10^8 \pm 0.30$

3. Materials and Methods

3.1. Materials and Reagents

Brasenia schreberi JF Gmel. was collected from Leibo Horse Lake, Sichuan Province, China. *B. schreberi* was dried at 40 °C then grounded into fine powder. The fat-soluble substances and pigment were removed using petroleum ether and 95% ethanol. After the powder was decolored and degreased, it was dried at 40 °C again and stored at −80 °C.

Petroleum ether, ascorbic acid (Vc), ethanol, congo red, sulfuric acid, phenol, and other reagents were purchased from Chengdu Kelon Chemical Co., Ltd. (Chengdu, China). Pectinase, glucoamylase, alpha-amylase, cellulase, and papainase were brought from Hefei Bomei Biotechnology Co., Ltd. (Hefei, China). Rose Bengal agar, MC medium, MRS medium, and Baird-Parker agar base were obtained from Guangdong Huankai Microorganism Technology Co., Ltd. (Guangzhou, China).

3.2. Polysaccharides from B. schreberi through Different Extraction Methods

Water extraction: Sample powder was mixed with distilled water in the ratio of 1:40 g/mL and set at 80 °C for 2 h with stirring. After repeating the extraction twice, the supernatant was collected.

Alkaline extraction: Sample powder was mixed with sodium hydroxide solution (0.2 M) in the ratio of 1:40 g/mL and set at 80 °C for 2 h with stirring. After repeating the extraction twice, the supernatant was collected.

Enzymatic extraction: Sample powder was mixed with distilled water in the ratio of 1:40 g/mL and cellulase was added into the mixture as 2%. Then, the solution was set at 60 °C for 2 h with stirring. After repeating the extraction twice, the supernatant was collected.

The supernatant was concentrated by a rotary evaporator and 4 volumes of ethanol were added in the concentration at 4 °C for 12 h. After centrifugation, the precipitation was dried at 50 °C and then redissolved in distilled water in the ratio of 4:1000 g/mL at 60 °C with the assistance of the ultrasonic disrupter for 15 min. Subsequently, the solution was centrifugated and the supernatant was concentrated and dialysed for 72 h before lyophilization to obtain 3 different polysaccharides water extractions: polysaccharides (W-BSP), alkaline extraction polysaccharides (A-BSP), and enzymatic extraction polysaccharides (E-BSP).

3.3. Assessment of Chemical Property

The total sugar content in the polysaccharides was detected using the phenol–sulfuric acid method [30]. The DNS method was used to detect the content of reducing sugar [52]. The protein content in polysaccharides was detected by the Coomassie brilliant blue method [53]. The iodine–potassium iodide reaction was used to detect whether the polysaccharide contained starch and the presence of phenolic hydroxyl groups in the polysaccharides was detected by the ferric chloride reaction [54,55].

3.4. Detection of Triple Helix Structure

The polysaccharides were dissolved in distilled water in the concentration of 1 g/L. 2 mL sample solution was mixed with 2 mL Congo red solution and a different concentration of sodium hydroxide solution. Then, the mixture was placed in the dark for 10 min before the absorbance at 300–600 nm was read.

3.5. Determination of the Antioxidant Ability

Three polysaccharides were dissolved in distilled water. Then, the scavenging ability of free radicals was performed according to the previous works [56].

3.6. Characterization of the Polysaccharide

Water-extracted polysaccharide (W-BSP) was dissolved in distilled water, then mixed with Sevage solution (chloroform: n-butanol = 4:1, *v*/*v*), and then shaken and centrifuged to remove the denatured proteins. Then, ADS-7 macroporous resin was added for decolorization. The final solution was eluted through the DEAE Cellulose DE-52 ion exchange cellulose column and Sephadex G-100 gel column, and then freeze-dried to obtain water-washed homogeneous polysaccharide.

The polysaccharide sample was mixed with KBr, grinded and pressed into slices, and the functional group was assessed using an infrared spectrum scan in the range of 4000–500 cm^{-1}. The monosaccharide composition of the polysaccharide was determined using high-performance liquid chromatography, and a gel chromatography-differential-multi-angle laser light scattering system (GPC-RI-MALLS) was used to determine the molecular weight. SEM electron microscopy was used to scan the surface of the polysaccharide and study its structure.

3.7. Preparation of Enzymolysis Polysaccharides

3.7.1. Detection of Enzymatic Hydrolysis Rate

W-BSP was dissolved in distilled water and the total polysaccharides and reducing sugar were determined as A and A0, respectively. The content of reducing sugar was taken as A1 after enzymatic hydrolysis, while the content of reducing sugar in the enzyme was recorded as A2. The enzymatic hydrolysis rate was calculated as the following formula:

$$\text{Enzymatic hydrolysis rate (\%)} = (A1 - A0 - A2)/A$$

3.7.2. Single Factor Experiment on Enzymatic Hydrolysis of Polysaccharides

The polysaccharide sample was dissolved in distilled water in the concentration of 4 mg/mL. Then, five enzymes (pectinase, glucoamylase, α-amylase, cellulase, and papain) were added in at different concentrations ranging from 10 U/mL to 70 U/mL. The pH of the mixture was adjusted by (0.1 M)-citric acid (0.05 M) to 3–6. And the solution was placed at 30–60 °C for 2–20 h. After the water bath, the enzymes were inactivated at 95 °C for 15 min. Then, the reaction mixture was centrifugated at 8000 rpm for 10 min. The content of reducing sugar in the supernatant was assessed and the enzymatic hydrolysis rate was calculated.

3.7.3. Response Surface Methodology for Enzymatic Hydrolysis of Polysaccharides

In order to optimize the enzymolysis process, Box–Behnken design (BBD) in response surface methodology (RSM) was applied to design a three-factor and three-level response surface experiment with the enzymolysis rate as the response value. Temperature, enzyme concentration, and pH were taken as independent variables. The BBD is shown in Supplementary Table S1. Due to the slow reaction velocity of enzymatic hydrolysis, the response surface optimization reaction time was set as 2 h in order to save time and cost. After the optimal conditions were obtained, the experiments were carried out under the optimized conditions of enzymolysis time (16 h) to obtain the maximum enzymolysis rate.

3.8. Cellular Assays

3.8.1. Cell Culture

HUVEC cells were obtained from the Institute of Cellular Sciences, Shanghai. Cells were maintained in DMEM medium containing 10% FBS at 37 °C in a 5% CO_2-incubator. When the cell fusion reached 80%, the cells were digested from the flask for the following assays.

3.8.2. Determination of Cell Toxicity

The HUVEC cells were seeded in a 96-well as the cell density was 8000 cells per well. After 16 h, different concentrations of the polysaccharides (resolved in PBS) were added into each well for a further 24 h incubation. Then, MTT (3-(4,5)-dimethylthiahiazo (-z-y1)-3,5-di- phenytetrazoliumromide) was transferred into the plate for 4 h and the formazan was resolved in DMSO. Finally, the absorbance of the plate was read by a microplate reader at 570 nm [57]. The cell viability was calculated according to the following formula:

$$\text{Cell viability (\%)} = 1 - A1/A2 \tag{1}$$

where A1 was the absorbance of the group with polysaccharides; and A2 was the absorbance of the group with PBS instead.

3.8.3. Determination of Cell Protection

The cell suspension was adjusted to 5000 cells per well. After 16 h, polysaccharides were added in the plate for 24 h. Then, the cells were exposed to H_2O_2 (100 μM) for 4 h. Finally, the cell viability was detected using the MTT method.

3.8.4. Determination of Cell Apoptosis and ROS Level

The cell density was adjusted to 20,000 cells per well in a 6-well plate. After 16 h incubation, polysaccharides were transferred into the plate for another 24 h. Then, the cells were exposed to H_2O_2 for 4 h. Subsequently, the cells were collected in 1.5 mL tubes and washed with cold PBS twice. AO/EB (100 μg/mL) were used to stain the cells for detecting cell apoptosis and DCFH-DA (2′,7′-Dichlorodihydrofluorescein diacetate, 10 μM, from Nanjing Jiancheng Biological Engineering Institute) was used to assess the ROS level through flow cytometry.

3.9. In Vivo Assays

3.9.1. *C. elegans* Maintenance

Wide type *C. elegans* was obtained from the *Caenorhabditis* Genetics Center (CGC). The worms were maintained in nematode growth media (NGM) plates with a layer of *Escherichia coli* OP50 as a food source in 20 °C. The synchronized worms were obtained using the bleach method [58]. And the polysaccharides were added into the medium and OP50.

3.9.2. Determination of Survival under Thermal Stress

Synchronized worms were laid down on the NGM plate until they grew to the L4 stage. Then, the worms were transferred to a new plate with or without polysaccharides for 48 h before they were exposed to 35 °C for 7 h. After thermal treatment, the worms were put back in 20 °C for 16 h to count the survival.

3.9.3. Determination of Reactive Oxygen Species (ROS)

After treatment with polysaccharides for 48 h, the worms were put in 35 °C for 4 h. Then, the worms were collected in a 1.5 mL tube and DCFH-DA (100 μM) was added in for 1 h in the dark. Subsequently the worms were washed with K medium 8 times and anesthetized by NaN_3 (40 mM). Then, the worms were transferred into 96-well plates with a density of 30 worms per well. The fluorescence density was read by a microplate reader and the worms' ROS level was observed by a fluorescence microscope.

3.9.4. Determination of Gene Expression

After 48 h treatment, the worms were exposed to hot condition for 4 h and then quickly collected and frozen in liquid nitrogen. The worms were then ground with a glass rod. The RNA was extracted according to the RNAeasy™ Animal RNA Extraction Kit (Cat: R0027, Beyotime Biotechnology, Shanghai, China). And the RNA was reversed to cDNA according to the PrimeScript™ RT reagent Kit with gDNA Eraser (Cat: RR047A, Takara Biomedical Technology (Beijing) Co., Ltd., Beijing, China). Then, the qPCR process was performed according to the iTaq™ Universal SYBR® Green Supermix (Cat: 1725124, Bio-Rad Laboratories, Inc., Hercules, CA, USA) in the CFX96 machine (Bio-Rad Laboratories, Inc., Hercules, CA, USA). The data were analyzed by $2^{-\Delta\Delta t}$ method [59].

3.10. Preparation of Polysaccharide Yogurt

The yogurt starter (*Lactobacillus bulgaricus*, *Streptococcus thermophilus*) was purchased from Angel Yeast Co., Ltd. (Hubei, China). The total polysaccharides sample was obtained after enzymatic hydrolysis and dried. The fermentation process was as follows: (A), polysaccharides were dissolved in pure milk at 60 °C; (B), sucrose was then added in for 6% (*w*/*v*); (C), the whole fermentation substrate was sterilized at 90 °C for 15 min; (D), the bacterial powder was inoculated at the concentration of 1.2 g/L after the mixture was cooled to 44 °C; (E), then the mixture was fermented at 42 °C for 5 h; and (F), finally, the yogurt product was obtained after ripening at 4 °C for 24 h.

3.10.1. Determination of Acidity in the Yogurt Fermentation Process

During the fermentation process, the fermented milk was taken out in a conical flask with 20 mL deionized water. Then, 2 mL phenolphthalein indicator was added in. After mixing, the mixture was titrated with sodium hydroxide (0.1 M) until slightly red, without fading, within 5 s. The titration acidity was defined as the volume of sodium hydroxide solution multiplied by 10.

3.10.2. Determination of the pH Value in the Yogurt Fermentation Process

During the fermentation process, the fermented milk was taken out to assess the pH value using a pH meter at 0 h, 1 h, 2 h, 3 h, 4 h, and 5 h.

3.10.3. Determination of the Water Holding Capacity of Yogurt

The yogurt was put in an empty tube (the weight of the tube was recorded as W1). And the total weight of the yogurt with the tube was W2. After centrifugation (3000 r/min, 10 min) and the removal of water, the residual matter was weighed as W3. The water holding capacity was calculated according to the following formula:

$$\text{Water holding capacity (\%)} = (W2 - W3)/(W2 - W1)$$

3.10.4. Measurement of Viscosity

A viscosimeter was applied to assess the viscosity of the yogurt. The post-ripened yogurt (10 g) was put in a tube, then the rotor F1 was stretched with the speed of 60 RPM. The data was recorded at 60 s, 90 s, and 120 s. The average value of the three times was taken as the final viscosity.

3.10.5. Determination of the Yogurt Textural Property

Texture profile analysis of the post-ripened yogurt was carried out using a texture analyzer with a 30 mm probe. The pre-test speed, the test speed, and the post-test speed all were 1 mm/s, and the measurement distance was 25 mm. The textural property was determined including Hardness (N), Elasticity (mm), Glueyness (N), Adhesivity (mj), and Adhesion elongation (mm).

3.10.6. Determination of Viable Count of Lactic Acid Bacteria

The yogurt was resolved in sterilized saline water in a shaken culture box. Then, the solution was diluted to 10^{-6}–10^{-9} four gradient dilution. 1 mL dilution was evenly spread on the MRS medium, and incubated at 36 °C in an anaerobic environment for 72 h. Finally, the colonies on the plates were recorded between 30 CFU/mL and 300 CFU/mL.

3.10.7. Sensory Evaluation Score

A total of 20 students and teachers majoring in food scored 100 points according to their taste, and the organizational structure, color, and flavor. The score details are shown in Supplementary Table S2.

3.11. Optimization of the Fermentation Conditions

3.11.1. Single Factor Assays

The polysaccharide sample was added into pure milk in the concentration of 0.05–0.25% (w/v), and sucrose was then added at 5–7% (w/v). After being fully stirred, different inoculation amounts of bacterial powder were subsequently transferred in (0.8–1.6 g/L). Then, the fermentation matrix was placed at 36–44 °C for 4.5–6.5 h to obtain the polysaccharide yogurt.

3.11.2. Optimization of Fermentation Conditions of Polysaccharide Yogurt by the Orthogonal Experiment

The fermentation temperature was set at 42 °C, the inoculation amount of bacteria was 1 g/L, and the sensory evaluation was directly applied as the evaluation index. An L9 (34) orthogonal experiment design was carried out on the three factors of polysaccharide addition, sucrose addition, and fermentation time. The design is shown in Supplementary Table S3.

3.11.3. Determination of the Indicators of the Polysaccharide Yogurt

Based on the results of orthogonal experiment, the polysaccharide yogurt was prepared using the optimized conditions. Then, the microorganisms were assessed including Lactic acid bacteria, *Escherichia coli*, *Staphylococcus aureus*, *Salmonella*, moulds, and yeasts according to the previous work [60]. In addition, the physical and chemical indexes were detected which involved acidity, protein content, fat, and water holding capacity [10,61].

3.12. Simulated Digestion of Polysaccharide Yogurt In Vitro

3.12.1. Production of Polysaccharide Yogurt

The polysaccharide was dissolved in pure milk at a concentration of 0.15% (w/v) at 60 °C. Subsequently, 6% sucrose was added to the solution, which was then sterilized at 90 °C for 20 min. After cooling to approximately 44 °C, 0.1 g of composite bacterial powder was incorporated, and the entire mixture was incubated at 42 °C for 5.5 h. Following this, the mixture underwent post-ripening at 4 °C, resulting in the production of polysaccharide yogurt.

3.12.2. Simulated Gastric Juice Digestion In Vitro

Pepsin (1%) was completely dissolved in hydrochloric acid solution at varying pH levels of 1.5, 2.5, and 3.5. Subsequently, the solution was filtered to acquire an in vitro simulated gastric juice. A specific quantity of polysaccharide yogurt was introduced into the simulated gastric juice, and the reaction was incubated at 37 °C on a shaking table operating at 100 r/min. The viable count of lactic acid bacteria was enumerated following 0, 1, 2, and 3 h of incubation.

3.12.3. Simulated Intestinal Fluid Digestion In Vitro

Ten grams of Trypsin was dissolved in distilled water to obtain solution A. Subsequently, 6.8 g of potassium dihydrogen phosphate was dissolved in 500 mL of water, and the pH was adjusted to 6.8 using 0.1 mol/L sodium hydroxide solution to obtain solution B.

Following this, the two solutions were combined and diluted to a final volume of 1000 mL with distilled water. The resultant solution was then filtered using a 0.22 μm microporous membrane to obtain simulated intestinal fluid for in vitro use. The polysaccharide yogurt was introduced into the simulated intestinal fluid and incubated at 37 °C on a shaking table rotating at 100 r/min. Viable counts of lactic acid bacteria were enumerated in the culture medium at 0, 1, 2, and 3 h.

3.13. Data Analysis

The data obtained in this work was analyzed by SPSS 22.0 software and were expressed as mean ± standard deviation. Graphpad Prism 8.0 was applied for drawing.

4. Conclusions

The hot water extraction method was found to be particularly suitable for extracting polysaccharides from *B. schreberi* (W-BSP) due to its high sugar content, excellent antioxidant activity, and triple helix structure. Following the enzymatic hydrolysis of W-BSP with pectinase, the resulting polysaccharides exhibited strong antioxidant properties both in vitro and in vivo. Notably, at a concentration of 250 μg/mL, the polysaccharides significantly increased the survival rate of *C. elegans* from 31.05% to 82.3%. Subsequently, these polysaccharides were introduced into yogurt fermentation, revealing that at an additional level of 0.2%, they demonstrated optimal water holding capacity, viscosity, elasticity, and sensory attributes. Furthermore, this addition promoted the proliferation of live lactic acid bacteria. In simulated in vitro gastrointestinal fluid digestion assays, the number of viable lactic acid bacteria in polysaccharide-enriched yogurt was higher compared to normal yogurt. These findings suggest that polysaccharides derived from *B. schreberi* represent a promising addition to yogurt, offering unique food functionality.

Supplementary Materials: The following supporting information can be downloaded at: https://www.mdpi.com/article/10.3390/molecules29010150/s1, Figure S1: The interaction of three factors on the enzymatic hydrolysis rate, including (A) the interaction of temperature and concentration of enzyme, (B) the interaction of pH value and temperature and (C) the interaction of pH value and concentration of enzyme.; Table S1: Box-Behnken design of the response surface method; Table S2: Sensory evaluation standards for polysaccharide yoghurt; Table S3: Factors and levels of orthogonal experiment of polysaccharide yoghurt; Table S4: Screening of degrading enzymes; Table S5: Results of orthogonal experiment.

Author Contributions: Methodology, R.F. and J.Z.; investigation, Y.W. and Y.Z.; data curation, Q.F.; writing—original draft preparation, Y.W.; supervision, W.Z. and Q.W. All authors have read and agreed to the published version of the manuscript.

Funding: This research was funded by the Innovation Research team of Yibin University grant number: No. 2018TD04 and the Innovation Research of Beijing Advanced Innovation Center for Food Nutrition and Human Health grant number: 20182017.

Institutional Review Board Statement: Not applicable.

Informed Consent Statement: Not applicable.

Data Availability Statement: Data are contained within the article and supplementary materials.

Conflicts of Interest: The authors declare no conflicts of interest.

References

1. Legault, J.; Perron, T.; Mshvildadze, V.; Girard-Lalancette, K.; Perron, S.; Laprise, C.; Sirois, P.; Pichette, A. Antioxidant and anti-inflammatory activities of quercetin 7-O-β-D-glucopyranoside from the leaves of *Brasenia schreberi*. *J. Med. Food* **2011**, *14*, 1127–1134. [CrossRef] [PubMed]
2. Hisayoshi, T.; Shinomura, M.; Yokokawa, K.; Kuze, I.; Konishi, A.; Kawaji, K.; Kodama, E.N.; Hata, K.; Takahashi, S.; Nirasawa, S.; et al. Inhibition of the DNA polymerase and RNase H activities of HIV-1 reverse transcriptase and HIV-1 replication by *Brasenia schreberi* (Junsai) and *Petasites japonicus* (Fuki) components. *J. Nat. Med.* **2015**, *69*, 432–440. [CrossRef] [PubMed]

3. Misaki, A.; Smith, F. Food plant polysaccharides, structure of the polysaccharide of the Japanese water-plant, *Brasenia schreberi* J. F Gmel. composition of the polysaccharide and isolation of 2-O-(β-D-glucopyranosyluronic acid)-D-mannose. *J. Agric. Food Chem* **1962**, *10*, 104–108. [CrossRef]
4. Kakuta, M.; Misaki, A. The polysaccharide of "Junsai (*Brasenia schreberi* J. F. Gmel)" Mucilage; Fragmentation Analysis by Successive Smith Degradations and Partial Acid Hydrolysis. *Agric. Biol. Chem.* **1979**, *43*, 1269–1276. [CrossRef]
5. Xiao, H.; Cai, X.; Fan, Y.; Luo, A. Antioxidant activity of water-soluble polysaccharides from *Brasenia schreberi*. *Pharmacogn. Mag* **2016**, *12*, 193–197. [CrossRef]
6. Kim, H.; Wang, Q.; Shoemaker, C.F.; Zhong, F.; Bartley, G.E.; Yokoyama, W.H. Polysaccharide gel coating of the leaves of *Brasenia schreberi* lowers plasma cholesterol in hamsters. *J. Tradit. Complement. Med.* **2015**, *5*, 56–61. [CrossRef]
7. Khursheed, R.; Singh, S.K.; Wadhwa, S.; Gulati, M.; Kapoor, B.; Jain, S.K.; Gowthamarajan, K.; Zacconi, F.; Chellappan, D.K.; Gupta, G.; et al. Development of mushroom polysaccharide and probiotics based solid self-nanoemulsifying drug delivery system loaded with curcumin and quercetin to improve their dissolution rate and permeability: State of the art. *Int. J. Biol. Macromol.* **2021**, *189*, 744–757. [CrossRef]
8. Mohanta, S.; Singh, S.K.; Kumar, B.; Gulati, M.; Kumar, R.; Yadav, A.K.; Wadhwa, S.; Jyoti, J.; Som, S.; Dua, K.; et al. Efficacy of co-administration of modified apple polysaccharide and probiotics in guar gum-Eudragit S100 based mesalamine mini tablets: A novel approach in treating ulcerative colitis. *Int. J. Biol. Macromol.* **2019**, *126*, 427–435. [CrossRef]
9. Khan, I.; Huang, G.; Li, X.; Leong, W.; Xia, W.; Hsiao, W.W.L. Mushroom polysaccharides from *Ganoderma lucidum* and *Poria cocos* reveal prebiotic functions. *J. Funct. Foods* **2018**, *41*, 191–201. [CrossRef]
10. Xu, K.; Guo, M.; Du, J.; Zhang, Z. Okra polysaccharide: Effect on the texture and microstructure of set yoghurt as a new natural stabilizer. *Int. J. Biol. Macromol.* **2019**, *133*, 117–126. [CrossRef]
11. Jonga, S.D.; Velde, F.V.D. Charge density of polysaccharide controls microstructure and large deformation properties of mixed gels. *Food Hydrocoll.* **2007**, *21*, 1172–1187. [CrossRef]
12. Li, S.; Nagendra, S.P. Effects of *Pleurotus eryngii* polysaccharides on bacterial growth, texture properties, proteolytic capacity, and angiotensin-I-converting enzyme-inhibitory activities of fermented milk. *J. Dairy Sci.* **2015**, *98*, 2949–2961. [CrossRef] [PubMed]
13. Saeedeh, S.; Shima, Y.; Alireza, F. Developing an innovative textural structure for semi-volume breads based on interaction of spray-dried yogurt powder and jujube polysaccharide. *Int. J. Biol. Macromol.* **2017**, *104*, 992–1002. [CrossRef]
14. Qiu, J.; Zhang, H.; Wang, Z. Ultrasonic degradation of Polysaccharides from *Auricularia auricula* and the antioxidant activity of their degradation products. *LWT-Food Sci. Technol.* **2019**, *113*, 108266. [CrossRef]
15. Zhang, H.; Cai, X.T.; Tian, Q.H.; Xiao, L.X.; Zeng, Z.; Cai, X.T.; Yan, J.Z.; Li, Q.Y. Microwave-assisted degradation of polysaccharide from *Polygonatum sibiricum* and antioxidant activity. *J. Food Sci.* **2019**, *84*, 754–761. [CrossRef]
16. Feng, S.; Luan, D.; Ning, K.; Shao, P.; Sun, P. Ultrafiltration isolation, hypoglycemic activity analysis and structural characterization of polysaccharides from *Brasenia schreberi*. *Int. J. Biol. Macromol.* **2019**, *135*, 141–151. [CrossRef]
17. Xu, X.; Yan, H.; Zhang, X. Structure and immuno-stimulating activities of a new heteropolysaccharide from *Lentinula edodes*. *J. Agric. Food Chem.* **2012**, *60*, 11560–11566. [CrossRef]
18. Ma, G.; Yang, W.; Mariga, A.M.; Fang, Y.; Ma, N.; Pei, F.; Hu, Q. Purification, characterization and antitumor activity of polysaccharides from *Pleurotus eryngii* residue. *Carbohydr. Polym.* **2014**, *114*, 297–305. [CrossRef]
19. Yang, B.; Wang, J.; Zhao, M.; Liu, Y.; Wang, W.; Jiang, Y. Identification of polysaccharides from pericarp tissues of litchi (*Litchi chinensis* Sonn.) fruit in relation to their antioxidant activities. *Carbohydr. Res.* **2006**, *341*, 634–638. [CrossRef]
20. Luo, Q.-L.; Tang, Z.-H.; Zhang, X.-F.; Zhong, O.-H.; Yao, S.-Z.; Wang, L.-S.; Lin, C.-W.; Luo, X. Chemical properties and antioxidant activity of a water-soluble polysaccharide from *Dendrobium officinale*. *Int. J. Biol. Macromol.* **2016**, *89*, 219–227. [CrossRef]
21. Sthijns, M.M.J.P.E.; Schiffers, P.M.; Janssen, G.M.; Lemmens, K.J.A.; Ides, B.; Vangrieken, P.; Bouwman, F.G.; Mariman, E.C.; Pader, I.; Arnér, E.S.J.; et al. Rutin protects against H_2O_2-triggered impaired relaxation of placental arterioles and induces Nrf2-mediated adaptation in Human Umbilical Vein Endothelial Cells exposed to oxidative stress. *Biochim. Biophys. Acta Gen. Subj.* **2017**, *1861*, 1177–1189. [CrossRef] [PubMed]
22. Sies, H. Hydrogen peroxide as a central redox signaling molecule in physiological oxidative stress: Oxidative eustress. *Redox Biol.* **2017**, *11*, 613–619. [CrossRef] [PubMed]
23. El-Kenawi, A.; Ruffell, B. Inflammation, ROS, and Mutagenesis. *Cancer Cell* **2017**, *32*, 727–729. [CrossRef] [PubMed]
24. Tafani, M.; Sansone, L.; Limana, F.; Arcangeli, T.; Santis, E.D.; Polese, M.; Fini, M.; Russo, M.A. The interplay of reactive oxygen species, hypoxia, inflammation, and sirtuins in cancer initiation and progression. *Oxidative Med. Cell. Longev.* **2016**, *2016*, 3907147. [CrossRef] [PubMed]
25. Schöttker, B.; Brenner, H.; Jansen, E.H.; Gardiner, J.; Peasey, A.; Kubínová, R.; Pająk, A.; Topor-Madry, R.; Tamosiunas, A.; Saum, K.-U.; et al. Evidence for the free radical/oxidative stress theory of ageing from the CHANCES consortium: A meta-analysis of individual participant data. *BMC Med.* **2015**, *13*, 300. [CrossRef] [PubMed]
26. Nobuhiko, K.; DeLano, F.A.; Schmid-Schönbein, G.W. Oxidative stress promotes endothelial cell apoptosis and loss of microvessels in the spontaneously hypertensive rats. *Arter. Thromb. Vasc. Biol.* **2005**, *25*, 2114–2121. [CrossRef]
27. An, R.; Wang, X.; Yang, L.; Zhang, J.; Wang, N.; Xu, F.; Hou, Y.; Zhang, H.; Zhang, L. Polystyrene microplastics cause granulosa cells apoptosis and fibrosis in ovary through oxidative stress in rats. *Toxicology* **2021**, *449*, 152665. [CrossRef]
28. Zhang, S.; Yi, X.; Su, X.; Jian, Z.; Cui, T.; Guo, S.; Gao, T.; Li, C.; Li, S.; Xiao, Q. *Ginkgo biloba* extract protects human melanocytes from H_2O_2-induced oxidative stress by activating Nrf2. *J. Cell. Mol. Med.* **2019**, *23*, 5193–5199. [CrossRef]

29. Jittiporn, K.; Moongkarndi, P.; Samer, J.; Kumphune, S.; Suvitayavat, W. Water extract of mangosteen suppresses H_2O_2-induced endothelial apoptosis by inhibiting oxidative stress. *J. Appl. Pharm. Sci.* **2019**, *9*, 10–16. [CrossRef]
30. Zhou, L.; Luo, S.; Li, J.; Zhou, Y.; Wang, X.; Kong, Q.; Chen, T.; Feng, S.; Yuan, M.; Ding, C. Optimization of the extraction of polysaccharides from the shells of *Camellia oleifera* and evaluation on the antioxidant potential in vitro and in vivo. *J. Funct. Foods* **2021**, *86*, 104678. [CrossRef]
31. Chen, T.; Luo, S.; Wang, X.; Zhou, Y.; Dai, Y.; Zhou, L.; Feng, S.; Yuan, M.; Ding, C. Polyphenols from *Blumea laciniata* extended the lifespan and enhanced resistance to stress in *Caenorhabditis elegans* via the insulin signaling pathway. *Antioxidants* **2021**, *10*, 1744. [CrossRef] [PubMed]
32. Wang, Y.; Luo, S.; Xu, Z.; Liu, L.; Feng, S.; Chen, T.; Zhou, L.; Yuan, M.; Huang, Y.; Ding, C. The potential antioxidant ability of hydroxytyrosol on *Caenorhabditis elegans* against oxidative damage via the insulin signaling pathway. *Arab. J. Chem.* **2021**, *14*, 103149. [CrossRef]
33. Zhao, D.; Yan, M.; Xu, H.; Liang, H.; Zhang, J.; Li, M.; Wang, C. Antioxidant and antiaging activity of fermented coix seed polysaccharides on *Caenorhabditis elegans*. *Nutrients* **2023**, *15*, 2474. [CrossRef] [PubMed]
34. Xiao, X.; Zhou, Y.; Tan, C.; Bai, J.; Zhu, Y.; Zhang, J.; Zhou, X.; Zhao, Y. Barley β-glucan resist oxidative stress of *Caenorhabditis elegans* via daf-2/daf-16 pathway. *Int. J. Biol. Macromol.* **2021**, *193*, 1021–1031. [CrossRef] [PubMed]
35. Arnab, M.; Seung, W.O.; Heidi, T. Worming pathways to and from DAF-16/FOXO. *Exp. Gerontol.* **2006**, *41*, 928–934. [CrossRef]
36. Blackwell, K.T.; Steinbaugh, M.J.; Hourihan, J.M.; Ewald, C.Y.; Isik, M. SKN-1/Nrf, stress responses, and aging in *Caenorhabditis elegans*. *Free Radic. Biol. Med.* **2015**, *88*, 290–301. [CrossRef] [PubMed]
37. Chatrawee, D.; Panthakarn, R.; Xiaojie, G.; Shaoxiong, Z.; Michael, W.; Tewin, T. *Glochidion zeylanicum* leaf extracts exhibit lifespan extending and oxidative stress resistance properties in *Caenorhabditis elegans* via DAF-16/FoxO and SKN-1/Nrf-2 signaling pathways. *Phytomedicine* **2019**, *64*, 153061. [CrossRef]
38. Li, N.; Li, Q.; He, X.; Gao, X.; Wu, L.; Xiao, M.; Cai, W.; Liu, B.; Zeng, F. Antioxidant and anti-aging activities of *Laminaria japonica* polysaccharide in *Caenorhabditis elegans* based on metabonomic analysis. *Int. J. Biol. Macromol.* **2022**, *221*, 346–354. [CrossRef]
39. Zhao, Y.; Zhi, L.; Wu, Q.; Yu, Y.; Sun, Q.; Wang, D. p38 MAPK-SKN-1/Nrf signaling cascade is required for intestinal barrier against graphene oxide toxicity in *Caenorhabditis elegans*. *Nanotoxicology* **2016**, *10*, 1469–1479. [CrossRef]
40. Han, Y.; Song, S.; Wu, H.; Zhang, J.; Ma, E. Antioxidant enzymes and their role in phoxim and carbaryl stress in *Caenorhabditis elegans*. *Pestic. Biochem. Physiol.* **2017**, *138*, 43–50. [CrossRef]
41. Chen, T.; Wang, X.; Yan, X.; Dai, Y.; Liang, T.; Zhou, L.; Feng, S.; Yuan, M.; Yang, H.; Ding, C. A novel selenium polysaccharide alleviates the manganese (Mn)-induced toxicity in Hep G2 cells and *Caenorhabditis elegans*. *Int. J. Mol. Sci.* **2022**, *23*, 4097. [CrossRef] [PubMed]
42. MatÉs, J.M.; Pérez-Gómez, C.; Castro, I.N.D. Antioxidant enzymes and human diseases. *Clin. Biochem.* **1999**, *32*, 595–603. [CrossRef]
43. Mataragas, M.; Dimitriou, V.; Skandamis, P.N.; Drosinos, E.H. Quantifying the spoilage and shelf-life of yoghurt with fruits. *Food Microbiol.* **2011**, *28*, 611–616. [CrossRef] [PubMed]
44. Gilbert, A.; Turgeon, S.L. Studying stirred yogurt microstructure and its correlation to physical properties: A review. *Food Hydrocoll.* **2021**, *121*, 106970. [CrossRef]
45. Ranasinghe, J.G.S.; Perera, W.T.R. Prevalence of *Lactobacillus bulgaricus* and *Streptococcus thermophilus* stability in commercially available yogurts in Sri lanka. *Asian J. Med. Sci.* **2016**, *7*, 97–101. [CrossRef]
46. Bracquart, P. An agar medium for the differential enumeration of *Streptococcus thermophilus* and *Lactobacillus bulgaricus* in yoghurt. *J. Appl. Bacteriol.* **1981**, *51*, 303–305. [CrossRef]
47. Sanders, M.E. Probiotics: Considerations for human health. *Nutr. Rev.* **2003**, *61*, 91–99. [CrossRef]
48. Hahn, C.; Sramek, M.; Nöbel, S.; Hinrichs, J. Post-processing of concentrated fermented milk: Influence of temperature and holding time on the formation of particle clusters. *Dairy Sci. Technol.* **2012**, *92*, 91–107. [CrossRef]
49. Miele, N.A.; Cabisidan, E.K.; Blaiotta, G.; Leone, S.; Masi, P.; Monaco, R.D.; Cavella, S. Rheological and sensory performance of a protein-based sweetener (MNEI), sucrose, and aspartame in yogurt. *J. Dairy Sci.* **2017**, *100*, 9539–9550. [CrossRef]
50. Salji, J.P.; Saadi, S.R.; Mashhadi, A. Shelf Life of Plain Liquid Yogurt Manufactured in Saudi Arabia. *J. Food Prot.* **1987**, *50*, 123–126. [CrossRef]
51. *GB4789.1-2016*; General Provisions for Food Microbiology Examination of National Food Safety Standards. Beijing, China, 2016. Available online: http://down.foodmate.net/standard/sort/3/50366.html (accessed on 23 June 2017). (In Chinese)
52. Ying, W.; Yuan, B.; Ji, Y.; Li, H. Hydrolysis of hemicellulose to produce fermentable monosaccharides by plasma acid. *Carbohydr. Polym.* **2013**, *97*, 518–522. [CrossRef]
53. Luo, S.; Jiang, X.; Jia, L.; Tan, C.; Li, M.; Yang, Q.; Du, Y.; Ding, C. In vivo and in vitro antioxidant activities of methanol extracts from olive leaves on *Caenorhabditis elegans*. *Molecules* **2019**, *24*, 704. [CrossRef] [PubMed]
54. Gaillard, B.D.E.; Thompson, N.S.; Morak, A.J. The interaction of polysaccharides with iodine. *Carbohydr. Res.* **1969**, *11*, 509–519. [CrossRef]
55. Xu, L.; Zhang, Y.; Wang, L. Structure characteristics of a water-soluble polysaccharide purified from dragon fruit (*Hylocereus undatus*) pulp. *Carbohydr. Polym.* **2016**, *146*, 224–230. [CrossRef] [PubMed]
56. Luo, S.; Li, J.; Zhou, Y.; Liu, L.; Feng, S.; Chen, T.; Zhou, L.; Ding, C. Evaluation on bioactivities of triterpenes from *Bergenia emeiensis*. *Arab. J. Chem.* **2021**, *14*, 103225. [CrossRef]

57. Zhou, L.; Luo, S.; Wang, X.; Zhou, Y.; Zhang, Y.; Zhu, S.; Chen, T.; Feng, S.; Yuan, M.; Ding, C. *Blumea laciniata* protected Hep G2 cells and *Caenorhabditis elegans* against acrylamide-induced toxicity via insulin/IGF-1 signaling pathway. *Food Chem. Toxicol.* **2021**, *158*, 112667. [CrossRef]
58. Porta-de-la-Riva, M.; Fontrodona, L.; Villanueva, A.; Cerón, J. Basic *Caenorhabditis elegans* methods: Synchronization and observation. *J. Vis. Exp.* **2012**, *64*, e2019. [CrossRef]
59. Livak, K.J.; Schmittgen, T.D. Analysis of relative gene expression data using real-time quantitative PCR and the $2^{-\Delta\Delta CT}$ Method. *Methods* **2001**, *25*, 402–408. [CrossRef]
60. Çakmakçı, S.; Cetin, B.; Turgut, T.; Gurses, M.; Erdoğan, A. Probiotic properties, sensory qualities, and storage stability of probiotic banana yogurts. *Turk. J. Vet. Anim. Sci.* **2012**, *36*, 231–237. [CrossRef]
61. Wang, J.; Liu, B.; Qi, Y.; Wu, D.; Liu, X.; Liu, C.; Gao, Y.; Shi, J.; Fang, L.; Min, W. Impact of *Auricularia cornea* var. Li polysaccharides on the physicochemical, textual, flavor, and antioxidant properties of set yogurt. *Int. J. Biol. Macromol.* **2022**, *206*, 148–158. [CrossRef]

molecules

MDPI

Article

Molecular Genomic Study of Inhibin Molecule Production through Granulosa Cell Gene Expression in Inhibin-Deficient Mice

Hira Sajjad Talpur [1,2], Zia ur Rehman [1,3,*], Mostafa Gouda [4,5,*], Aixing Liang [1], Iqra Bano [2], Mir Sajjad Hussain [2], FarmanUllah FarmanUllah [1] and Liguo Yang [1,*]

1 National Center for International Research on Animal Genetics, Breeding and Reproduction (NCIRAGBR), Huazhong Agricultural University, Ministry of Science and Technology of the People's Republic of China, Wuhan 430070, China
2 Department of Animal Breeding and Genetics, Sindh Agriculture University, Tandojam 70060, Sindh, Pakistan
3 College of Veterinary Sciences, Faculty of Animal Husbandry and Veterinary Sciences, University of Agriculture, Peshawar 25120, Khyber Pakhtunkhwa, Pakistan
4 College of Biosystems Engineering and Food Science, Zhejiang University, Hangzhou 310058, China
5 Department of Nutrition & Food Science, National Research Centre, Dokki, Giza 12622, Egypt
* Correspondence: drzia80@aup.edu.pk (Z.u.R.); Mostafa-gouda@zju.edu.cn or goudarowing@yahoo.com (M.G.); ylg@mail.hzau.edu.cn (L.Y.)

Abstract: Inhibin is a molecule that belongs to peptide hormones and is excreted through pituitary gonadotropins stimulation action on the granulosa cells of the ovaries. However, the differential regulation of inhibin and follicle-stimulating hormone (FSH) on granulosa cell tumor growth in mice inhibin-deficient females is not yet well understood. The objective of this study was to evaluate the role of inhibin and FSH on the granulosa cells of ovarian follicles at the premature antral stage. This study stimulated immature wild-type (WT) and Inhibin-α knockout (Inha−/−) female mice with human chorionic gonadotropin (hCG) and examined hCG-induced gene expression changes in granulosa cells. Also, screening of differentially expressed genes (DEGs) was performed in the two groups under study. In addition, related modules to external traits and key gene drivers were determined through Weighted Gene Co-Expression Network Analysis (WGCNA) algorithm. The results identified a number of 1074 and 931 DEGs and 343 overlapping DEGs (ODEGs) were shared in the two groups. Some 341 ODEGs had high relevance and consistent expression direction, with a significant correlation coefficient (r^2 = 0.9145). Additionally, the gene co-expression network of selected 153 genes showed 122 nodes enriched to 21 GO biological processes (BP) and reproduction and 3 genes related to genomic pathways. By using principal component analysis (PCA), the 14 genes in the regulatory network were fixed and the cumulative proportion of fitted top three principal components was 94.64%. In conclusion, this study revealed the novelty of using ODEGs for investigating the inhibin and FSH hormone pathways that might open the way toward gene therapy for granulosa cell tumors. Also, these genes could be used as biomarkers for tracking the changes in inhibin and FSH hormone from the changes in the nutrition pattern.

Keywords: FSH; DEGs; granulosa cells; hCG; inhibin; Inha

1. Introduction

Gonadotropins are glycoprotein hormones produced in the pituitary by gonadotropic cells the lightest copious ovarian cells and regulate ovarian follicle development [1,2]. These peptide hormones are also known for regulating ovarian and testicular function and are essential for normal growth, sexual development, and reproduction. Also, gonadotropins include some essential hormones like inhibin, FSH, and luteinizing hormone (LH) [2,3]. For instance, inhibin is released from the ovarian granulosa cell. It is a heterodimeric glycoprotein that consists of α-subunit linked with a β-hormone [4]. It is a member of the transforming growth factor-β superfamily that stimulate the release of FSH from the pituitary cells. Moreover, inhibin plays an important role in the FSH feedback secretion regulation during puberty in females [5]. In which, its secretion changes during female puberty have disclose correlation with adulthood chronic diseases like diabetes type 2 and heart disease which are generally considered among the common malnutrition diseases [6]. In addition, the inhibin hormone is essential as a diagnostic marker for ovarian cancer [7,8]. For instance, Rathore, et al. [9] mentioned that Inhibin deficiency in mice leads to the growth of gonadal sex cord-stromal tumors. In such cases, the mice mostly died at the age of 28 days due to testicular and ovarian tumors with cachexia-related signs [10–12]. A genetic method assured that gonadotropins play a key modifier role for gonadal sex cord-stromal tumor growth in inhibin-deficient mice. In that study, Nagaraja, et al. [13] reported that inhibin genetically interacts with multiple factors that influence testicular and ovarian growth and differentiation, including the pituitary gonadotropins [14]. Furthermore, they found a complex interplay among inhibins, gonadotropins, and ovarian cancer. Thus, the tracking of inhibin and FSH hormones could enhance the diagnostic performance of the health-related disease, not just ovarian female disease.

Furthermore, it is known that granulosa cells (GCs) are somatic cells of the sex rope that is mainly related to an embryonic female gamete identified to be an oocyte that exists in the animals' ovary [15]. GCs show various phenotypes in the follicle, reliant on their position. Additionally, GCs gene expression analysis is very important to understand its functional mechanism that is related to animal growth. Hence gene association networks are very essential for expressing the relationship patterns between genes transversely microarray data, where the weighted gene co-expression network analysis (WGCNA) algorithm is a necessary tool to determine the relationship patterns among the genes microarray data [16], for which, the WGCNA algorithm is frequently used to understand the genes' molecular processes and identification of interrelated genes and modules [17]. Therefore, it can show the co-expression structure and cluster the expression data into modules of conserved function that allow one to detect patterns of gene connectivity that can be aligned with behavioral and physiological phenotypes [18]. Meanwhile, PCA is a multivariate statistical procedure that uses an orthogonal transformation to convert a set of observations of possibly correlated variables into a set of values of linearly uncorrelated variables called principal components [19–21].

Therefore, this study aimed to investigate the variations in hCG-induced gene expression in WT and Inha−/− granulosa cells. It provided a novel reference for the pathway mechanisms. Initially, differentially expressed genes (DEGs) in WT/Inha−/−with and without human chorionic gonadotropin (hCG) stimulation samples were identified. Then the overlapping genes as the characteristic genes in hCG stimulation and Inha−/− female mice granulosa cells were selected and further investigated. For this process, the WGCNA algorithm was used to identify the interrelated gene patterns. Besides, PCA was used to examine the important regulated genes. The examined outcome recommended that Inhibin α knockout and hCG stimulation can down-regulate JUP expression, and up-regulate Psmc3ip expression. These genes could be used as a marker for the early prediction of granulosa cell carcinomas.

2. Results

2.1. Hierarchical Clustering and Comparison Analysis of Selected DEGs in Different Groups

We extracted DEG expressions from two groups and drew hierarchical clustering heatmaps, as shown in Figure 1. The experimental samples in each group were divided into two separate parts, indicating that DEGs had obviously different expression patterns in each group (gene expression matrix can be found in Table S1, Supplementary Materials).

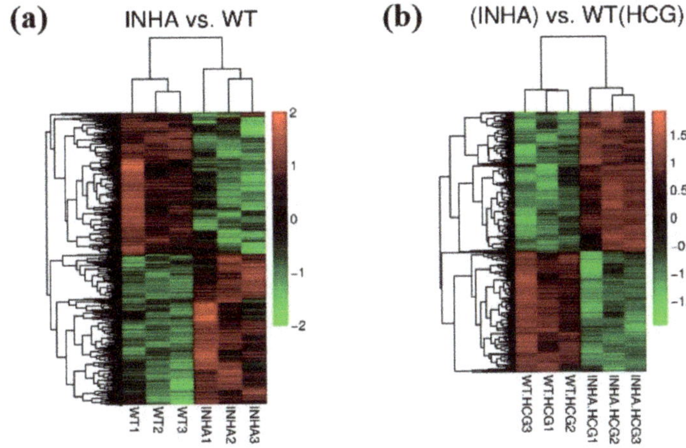

Figure 1. Hierarchical clustering heatmap of DEGs in group Inha−/− vs. WT (**a**) and Inha−/− (hCG) vs. WT(hCG) (**b**).

2.2. GO and KEGG Pathway Enrichment Analysis for the ODEGs

Figure 2a shows the Venn diagram of DEGs in groups Inha−/− vs. WT and Inha−/− (hCG) vs. WT(hCG). In which, 343 overlapped genes were found with a total of 25 (10 BP, 10 CC, 5 MF) significant related GO annotations for 343 overlapped genes that listed in Table 1. Also, the hierarchical clustering heatmap of ODEGs showed a significant negative correlation of the down-regulated Inha DEGs compared to the WT group (Figure 2b,c). What's more, among the 343 shared DEGs, 341 had a consistent expression direction and high relevance, with a significant correlation coefficient of 0.9145 ($p < 0.000001$). The shared 343 DEGs were differentially expressed in both Inha−/− vs. WT group and Inha−/− (hCG) vs. WT(hCG) group (Table S1). Meanwhile, the GO BPs were significantly related to the cell cycle process (GO: 0022402), which had the most enrichment significance (p-value < 0.0001). Also, 17 genes were involved, such as CDC6, KIFC1, MKI67, DSN1, NUF2. Besides, according to KEGG enrichment analysis, ODEGs were significantly enriched in eight pathways (RFC5, PRIM1, RPA2, RFC4, LIG1, POLD2, POLE, MCM2) participating in the most significant related pathway: DNA replication (mmu03030). Using the ggplot2 package in R, significant related GO and KEGG pathways annotations were displayed in Figure 3.

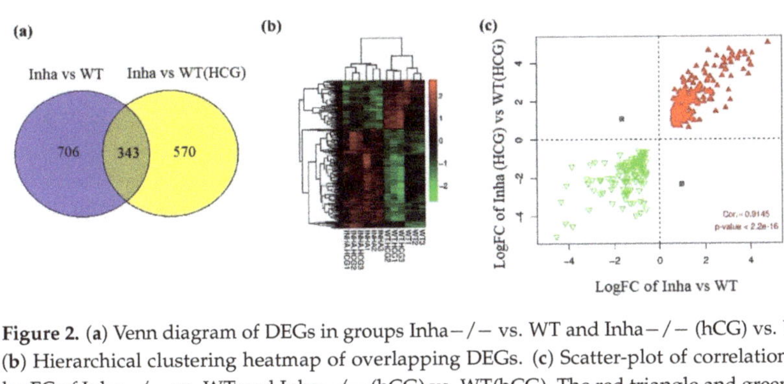

Figure 2. (**a**) Venn diagram of DEGs in groups Inha−/− vs. WT and Inha−/− (hCG) vs. WT(hCG). (**b**) Hierarchical clustering heatmap of overlapping DEGs. (**c**) Scatter-plot of correlation between logFC of Inha−/− vs. WT and Inha−/− (hCG) vs. WT(hCG). The red triangle and green inverted triangle refer to up and down-regulated DEGs in both Inha−/− vs. WT and Inha−/− (hCG) vs. WT(hCG).

Table 1. Enriched GOs and KEGG pathways for overlapped genes.

Category	Term	Count	*p*-Value
Biology Process	GO:0022402~cell cycle process	17	4.05×10^{-6}
	GO:0006259~DNA metabolic process	17	8.56×10^{-4}
	GO:0022403~cell cycle phase	16	1.79×10^{-4}
	GO:0000279~M phase	15	1.31×10^{-4}
	GO:0010033~response to organic substance	15	2.53×10^{-2}
	GO:0006260~DNA replication	13	4.14×10^{-6}
	GO:0051301~cell division	11	1.21×10^{-2}
	GO:0001568~blood vessel development	10	1.37×10^{-2}
	GO:0001944~vasculature development	10	1.58×10^{-2}
	GO:0001501~skeletal system development	10	3.33×10^{-2}
Cellular Component	GO:0005578~proteinaceous extracellular matrix	19	5.11×10^{-6}
	GO:0005615~extracellular space	19	4.03×10^{-3}
	GO:0044427~chromosomal part	15	1.49×10^{-3}
	GO:0005694~chromosome	15	7.06×10^{-3}
	GO:0044454~nuclear chromosome part	8	2.25×10^{-3}
	GO:0000228~nuclear chromosome	8	5.51×10^{-3}
	GO:0005657~replication fork	6	5.21×10^{-5}
	GO:0000793~condensed chromosome	6	4.05×10^{-2}
	GO:0042383~sarcolemma	5	9.21×10^{-3}
	GO:0030018~Z disc	4	4.69×10^{-2}
Molecular Function	GO:0019838~growth factor binding	6	4.09×10^{-3}
	GO:0008094~DNA-dependent ATPase activity	4	1.80×10^{-2}
	GO:0016875~ligase activity, forming carbon-oxygen bonds	4	3.32×10^{-2}
	GO:0004812~aminoacyl-tRNA ligase activity	4	3.32×10^{-2}
	GO:0005520~insulin-like growth factor binding	3	4.08×10^{-2}
KEGG Pathway	mmu03030:DNA replication	8	1.35×10^{-6}
	mmu03430:Mismatch repair	5	4.27×10^{-4}
	mmu04512:ECM-receptor interaction	8	4.34×10^{-4}
	mmu03420:Nucleotide excision repair	6	6.87×10^{-4}
	mmu03440:Homologous recombination	4	9.92×10^{-3}
	mmu04510:Focal adhesion	9	1.73×10^{-3}
	mmu03410:Base excision repair	4	2.87×10^{-2}
	mmu00230:Purine metabolism	7	4.71×10^{-2}

2.3. *Physiological Phenotypes R Modules and Genes Identification Based on WGCNA*

For clustering the ODEGs based on the physiological phenotypes, the expression data were processed by square root transformation and used to infer co-expression gene network modules with the WGCNA network construction and the module detection method. Firstly, the distances among all the samples were studied to eliminate discrete samples with no discrete samples to be removed (Figure 4a). In which, a high affinity between Inha and Inha(hCG) was found by the clustering tree. Then a proper power-law coefficient was selected using the soft-threshold method (Figure 4b). Through this model the selected soft-threshold (X-axis) was 18 when the scale-free topology model fit was signed at correlation coefficient (R^2) = 0.8 (Y-axis). Then, a dynamic hierarchical tree cut algorithm was used to detect the co-expression modules, and a total of six related modules were found (Figure 4c). Moreover, the R^2 between the physiological phenotypes and each module had a very high correlation coefficient (over 0.8) with the physiological phenotypes (Figure 4d and Table 2). Thus, 153 genes in the top three modules (blue, green, and brown) in total were selected as representative ODEGs for further analysis based on their significant physiological phenotypes.

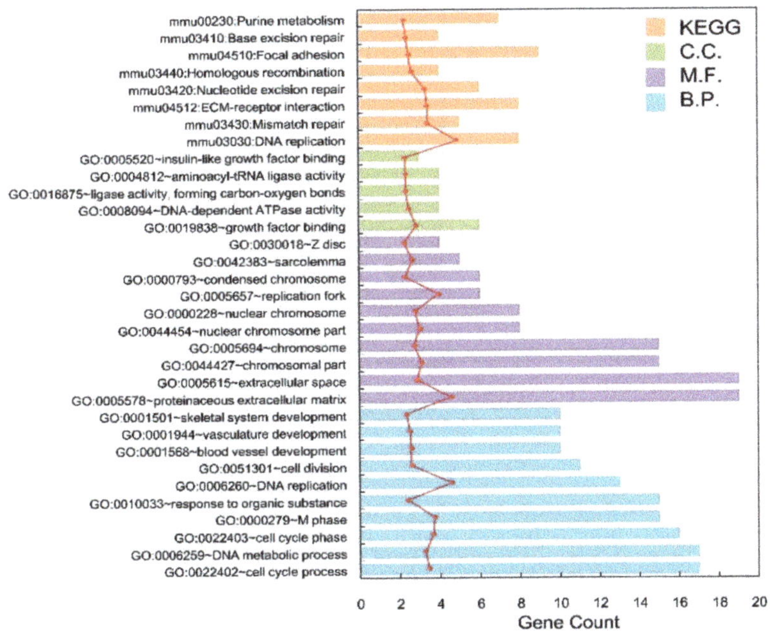

Figure 3. The histogram of the category of enriched GO terms and KEGG pathways for the overlapped DEGs. The horizontal axis represents the number of blue, purple, green, and orange mean Biology Process, Cellular Component, Molecular Function, and pathways, respectively; the red dot curve means –log10 (*p*-value).

Table 2. Correlation between physiological phenotypes and each module genes.

Color	Gene Count	Correlation Coefficient (R^2)
blue	65	0.9218561
green	28	0.9203381
brown	60	0.8912573
turquoise	81	0.8894315
yellow	51	0.889348
grey	58	0.8815949

2.4. Co-Expression Network Construction

The Co-expression Network had 153 DEGs in blue, green, and brown modules, as well as an expression correlation coefficient from the WGCNA algorithm (expression correlation matrix was shown in Table S2). Also, the selected gene pair was based on the expression correlation coefficient >0.8 as shown in Figure 5. The co-expression network included 122 nodes in total. These nodes had 35 down-regulated genes (14 blue, 17 brown, and 4 green) and 87 up-regulated genes (30 blue, 36 brown, and 21 green genes) with 410 edges (129 negative coefficient connections and 281 positive coefficient connections) (Tables 2 and S3).

Moreover, GO and KEGG pathway enrichment analysis showed that a total of 21 significantly related GO BPs and 3 KEGG pathways were found for the DEGs in the co-expression network. DEGs in the co-expression network were significantly (p-value < 0.01) related to the cell cycle BPs and participated in ECM-receptor interaction (mmu04512), Focal adhesion (mmu04510), and DNA replication (mmu03030) pathways (Figure 6 and Table 3).

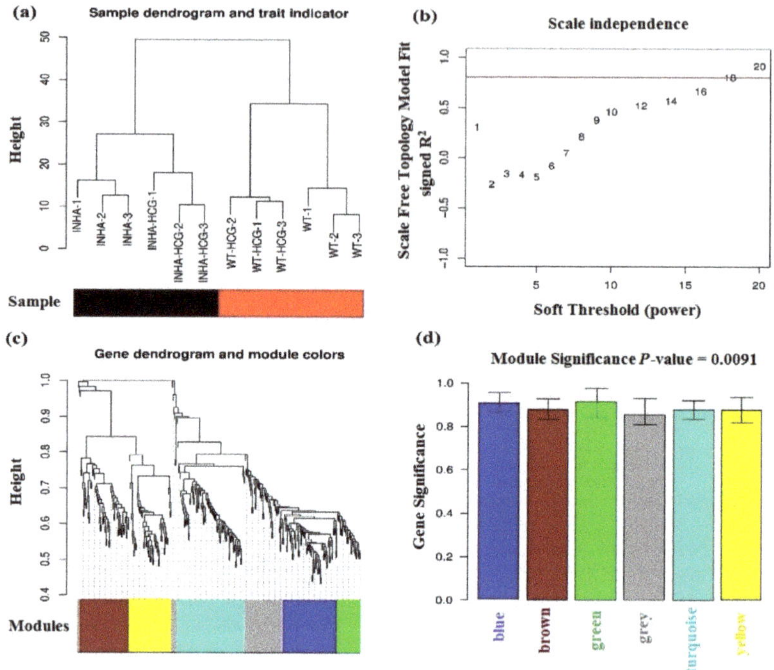

Figure 4. (**a**) Sample clustering tree. The Black and red bars mean different types of samples. (**b**) power-law coefficient parameter plot X-axis means soft-threshold, Y-axis means scale-free topology model fit signed R-square. (**c**) Modules clustering tree, different colors in the bottom mean different modules. (**d**) Module bar plot, X-axis means different modules, Y-axis means the significance of genes in each color module based on their different physiological phenotypes.

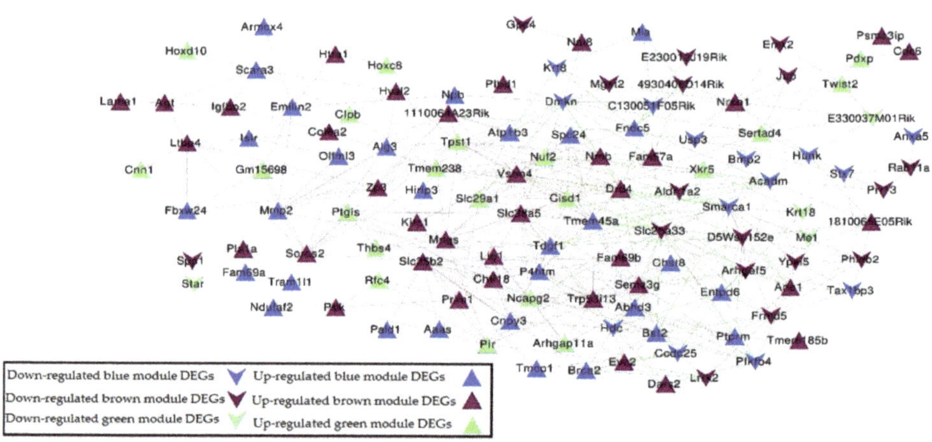

Figure 5. Gene co-expression network based on 153 DEGs in blue, green, and brown modules. Triangle and inverted triangle refer to up and down-regulated DEGs; blue, green, and brown nodes mean genes from the corresponding colored module. Redline connections mean a positive correlation coefficient, and green line connections mean a negative correlation coefficient.

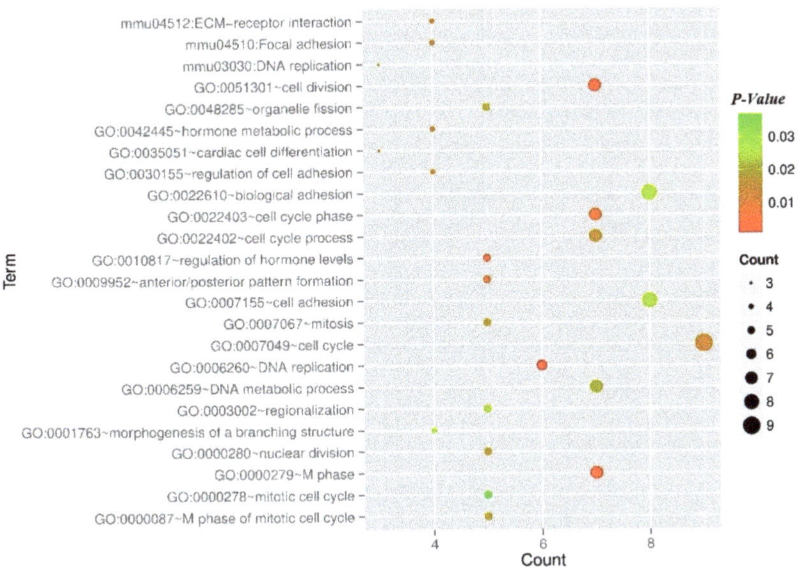

Figure 6. The scatterplot of the category of enriched GO terms and KEGG pathways for the DEGs in the co-expression network. Node size means gene count, color means *p*-value.

Table 3. Enriched GOs and KEGG pathways for co-expression network genes.

Parameter	Term	Count	*p*-Value
Biology Process	GO:0007049~cell cycle	9	0.012955
	GO:0007155~cell adhesion	8	0.02503
	GO:0022610~biological adhesion	8	0.025243
	GO:0051301~cell division	7	0.003211
	GO:0000279~M phase	7	0.003325
	GO:0022403~cell cycle phase	7	0.00678
	GO:0022402~cell cycle process	7	0.015593
	GO:0006259~DNA metabolic process	7	0.021132
	GO:0006260~DNA replication	6	0.001106
	GO:0010817~regulation of hormone levels	5	0.004264
	GO:0009952~anterior/posterior pattern formation	5	0.007972
	GO:0000280~nuclear division	5	0.016588
	GO:0007067~mitosis	5	0.016588
	GO:0000087~M phase of mitotic cell cycle	5	0.017769
	GO:0048285~organelle fission	5	0.018688
	GO:0003002~regionalization	5	0.024455
	GO:0000278~mitotic cell cycle	5	0.03702
	GO:0042445~hormone metabolic process	4	0.011174
	GO:0030155~regulation of cell adhesion	4	0.012565
	GO:0001763~morphogenesis of a branching structure	4	0.026619
	GO:0035051~cardiac cell differentiation	3	0.012461
KEGG pathway	mmu04512:ECM-receptor interaction	4	0.010701
	mmu04510:Focal adhesion	4	0.009668
	mmu03030:DNA replication	3	0.015987

2.5. miRNA-DEGs-TF Regulatory Network Construction

MicroRNAs (miRNAs) assume a pivotal role in controlling inborn and versatile immunity in humans and animals [22,23]. A total number of eight miRNAs and seven

TFs that have the potential to regulate ODEGs in the co-expression network are listed in Tables 4 and 5, respectively. Integration regulatory relationships were identified and constructed as a miRNA-DEGs-TF regulatory network between miRNA and DEGs, TFs and DEGs (Figure 7). In that network, 29 nodes (8 miRNAs, 7 TFs, and 14 DEGs), 7 down (1 blue, 6 brown) and 7 up-regulated (3 blue, 2 brown, and 2 green genes) and 56 edges (20 miRNA-DEGs regulation and 36 TFs-DEGs) have been confirmed (Table S4).

Table 4. Related miRNAs list.

miRNA	ID	*p*-Value	FDR
mmu_TGCCTTA,MIR-124A	DB_ID:590	9.65×10^{-3}	0.0014
mmu_GTGACTT,MIR-224	DB_ID:524	0.0002	0.0014
mmu_CTCTGGA,MIR-520A	DB_ID:484	0.0036	0.0126
mmu_ACCAAAG,MIR-9	DB_ID:588	0.0029	0.0126
mmu_ACTGAAA,MIR-30A	DB_ID:464	0.0065	0.0182
mmu_CTGAGCC,MIR-24	DB_ID:539	0.0107	0.0194
mmu_AACTGGA,MIR-145	DB_ID:614	0.0101	0.0194
mmu_AAGCACT,MIR-520F	DB_ID:615	0.0111	0.0194

Table 5. Related TFs list.

TF	ID	*p*-Value	FDR
PAX4	DB_ID:1830	8.59×10^{-6}	2.58×10^{-5}
MAZ	DB_ID:1815	7.72×10^{-6}	2.58×10^{-5}
MYC	DB_ID:1819	4.99×10^{-6}	2.58×10^{-5}
NFAT	DB_ID:1822	1.40×10^{-5}	3.15×10^{-5}
FOXO4	DB_ID:1801	3.72×10^{-5}	6.70×10^{-5}
SP1	DB_ID:1837	2.00×10^{-4}	3.00×10^{-4}
LEF1	DB_ID:1813	4.00×10^{-4}	4.00×10^{-4}

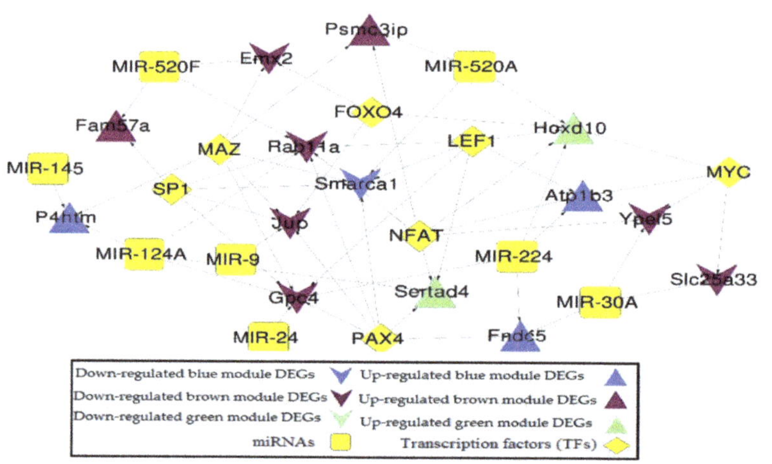

Figure 7. miRNA-DEG-TF regulatory network of DEGs in the co-expression network. Triangle and inverted triangles refer to up and down-regulated DEGs; blue, green, and brown nodes mean genes from the corresponding colored module. Yellow square and diamond mean miRNAs and TFs.

2.6. PCA for Genes in Regulatory Network

To further refine the important genes, the PCA algorithm defined 14 genes in the regulatory network. The cumulative proportion of fitted top 3 principal components accounts for 94.64% of the total variance, which means that they can effectively describe the vast majority of input gene variables [20]. In general, the cumulative contribution rate of more than 80% is considered to have caught most of the input variable information. Figure 8 showed the differences in the top 3PCs between the two groups where significant differences ($p < 0.01$) were observed among the four different groups. Also, the gene contributions to the PCs were listed in Table 6, where there were 10 genes whose contribution rate (absolute value) was over 0.9.

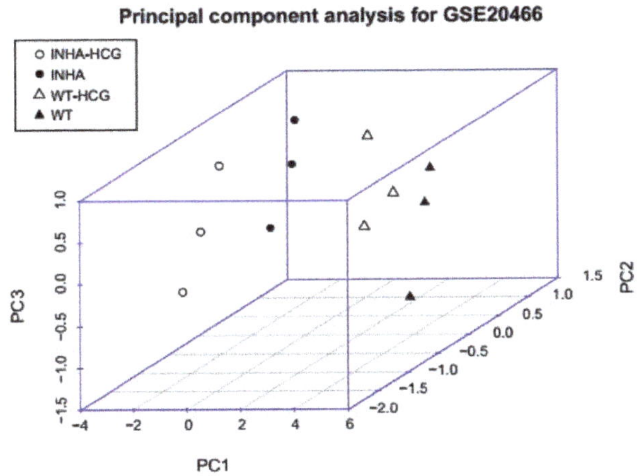

Figure 8. Three-dimensional distribution of samples based on PC1, PC2, and PC3. Black solid points and hollow points mean Inha−/− with and without hCG samples, and solid and hollow triangles mean WT with and without hCG samples.

Table 6. Gene contributions to PC1-3.

Gene	Contribution to PC1-3
Fndc5	0.97
Sertad4	0.97
Atp1b3	0.96
Fam57a	0.95
P4htm	0.89
Hoxd10	0.85
Psmc3ip	0.85
Rab11a	−0.85
Ypel5	−0.9
Emx2	−0.91
Jup	−0.92
Gpc4	−0.93
Slc25a33	−0.94
Smarcal	−0.94

3. Discussion

It is shown that Inhibin-α plays an important role in follicular development, oocyte development, cell differentiation, and finally reproduction. The Inhibin-α knockout/down female mouse can develop ovarian cancer and the LH and FSH may play a crucial role in GCs tumor development [24,25]. This study aimed to examine hCG-induced gene expression changes in different types of granulosa cells (WT and Inha−/− types). Also, it provided an important reference for the pathway mechanisms by showing that DEGs were different in WT/Inha−/− before and after hCG stimulation. The ODEGs were used as characteristic genes in hCG stimulation and Inhibin α knockout (Inha−/−) female mice granulosa cells. This observation is in agreement with FarmanUllah, Liang, Khan,

Salim, Rehman, Khan, Talpur, Schreurs, Gouda, Khan and Shujun [22] who mentioned that ODEGs can effectively work as biomarkers for immune-related tumors. In which, 341 DEGs had high relevance, with a significant correlation coefficient ($p < 0.000001$) in both Inha−/− vs. WT group and Inha−/− (hCG) vs. WT(hCG) group which means that they could be used as characteristic genes in hCG stimulation and Inhibin α knockout (Inha−/−) female mice granulosa cells. Therefore, this study demonstrated for the first time that hCG induces the granulosa cells to excrete Inha through stimulating Fndc5, Sertad4, JUP, and Psmc3ip genes. In which, quantitative reverse transcriptase-polymerase chain reaction (qRT-PCR) on selected gene expression changes were observed in the gene array analysis verified the most important ODEGs of the knockout mice. Vasilache, et al. [26] mentioned that qRT-PCR microarray combined with modeling is an effective technique to detect the knockout mice's important DEGs.

According to the analysis of KEGG pathway annotations, ODEGs were significantly enriched in eight biological pathways: DNA replication, Focal adhesion, and purine metabolism pathways which significantly enriched the GO term. In the biological process category, the genes were mainly enriched in GO terms associated with extracellular matrix and axon [27], for which, the R^2 between the physiological phenotypes and each KEGG module for DEGs had a very high correlation coefficient (>0.8). Chen, et al. [28] mentioned that a positive regulator of the steroidogenesis pathway of FSH is essential for the granulosa cell proliferation, death, and differentiation in almost all cell types. The WGCNA algorithm was used to detect related modules and genes, significantly related to eight miRNAs; whereas, the seven TFs and regulatory networks were utilized to get regulated DEGs [29]. Finally, PCA differentiated the four groups under study to determine the importance of 14 regulated genes. Also, the top 3PCs between the two groups were significantly different ($p < 0.01$) among the four groups under study. Among them, the JUP gene was significantly related to cell adhesion (GO: 0007155), DNA metabolic process (GO:0006259), and biological adhesion (GO: 0022610), while Psmc3ip participated in the cell cycle (GO: 0022402) and M phase (GO: 0000279), and they all belonged to the brown module of the physiological phenotypes of WGCNA results, so they had close expression relationship. Chen, et al. [30] reported that the JUP form is a member of the catenin family that can affect various processes such as proliferation, migration, and differentiation by mediating cellular adhesion. Thus, the mutation in its gene is associated with several gene-related diseases. [30]. In addition, Psmc3ip (also known as GT198) is used as a unique tumor marker suppressor gene for the mutant cells in ovarian cancer. Psmc3ip protein has been shown as a steroid hormone receptor regulator and also as a crucial factor in DNA repair [31]. Thus, studying such genes could facilitate the complex mission of dealing with ovarian cancer.

Additionally, the result suggested that Inhibin α knockout and hCG stimulation can down-regulate the expression of JUP and up-regulate Psmc3ip. In which, the co-expression Network had 153 DEGs expression correlation coefficient from WGCNA algorithm with $R^2 > 0.8$ with 35 down-regulated genes and 87 up-regulated genes. Moreover, JUP forms distinct complexes with cadherins and desmosomal cadherins through an amino acid motif called the armadillo repeat, which can affect the diverse processes and modulate the function of extracellular ligands [32,33]. It also showed that JUP and Psmc3ip genes had close relationships in both expression patterns and functions in the Inha−/− hCG stimulation female mice granulosa cells. Similarly, the activity of the Psmc3ip gene is revealed to have a crucial role in ovarian dysgenesis and male fertility in mammalians [34,35]. Thus, the data analysis detected DEGs and relevant biological functions after the knockdown of the Inha and associated gene expression for further research guidance in mammalian reproduction.

4. Materials and Methods

4.1. Experimental Animals

A number of 100 specific-pathogen-free (SPF) mice were grouped (25 mice WT, 25 mice Inha−/−, 25 mice WT (hCG), and 25 mice Inha−/−(hCG)) according to Hofland, et al. [36]. In which, 21 to 23-day-old WT and Inha−/− female mice were injected with 5 IU hCG for

6 h to stimulate hCG groups and granulosa cells with and without hCG stimulation which were collected from 2 genotypes (WT and Inha−/−) according to National Institutes of Health (NIH) Guidelines for the Care and use of Laboratory Animals, USA (Approval ID: SCXK Hubei 20080005).

4.2. Data and Experimental Design

A schematic diagram of the overall research procedure for data analysis is shown in Figure 9. Dataset and description.

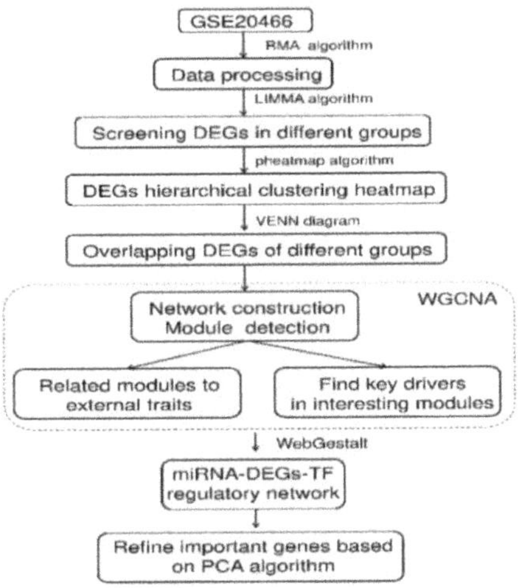

Figure 9. Schematic illustration of the analysis strategy.

The target gene expression profiles were downloaded from NCBI Gene Expression Omnibus (GEO; http://www.ncbi.nlm.nih.gov/geo/ accessed on 1 February 2011) through accession number GSE20466 (Platform: GPL1261 [Mouse430_2] Affymetrix Mouse Genome 430 2.0 Array), which contained 12 samples in total [37,38].

4.3. Data Reprocessing and Differentially Expressed Genes (DEGs) Screening

The main objective of this part was to initially normalize the datasets' differences and functions. The data before and normalization were shown in Figure 10, and the detailed normalized gene expression data can be found in Table S5. Only those genes meeting FDR < 0.05 and |log2 FC (fold change)| > 1 were chosen as DEGs from each group. In Inha−/− vs. WT and Inha−/− with hCG vs. WT groups only 1074 and 931 DEGs were identified based on the cut-off criteria and showed in volcano plots for WT (Figure 11A), and Inha−/− (Figure 11B) respectively. The list of DEGs could be found in Table S6 (Supplementary Materials).

Raw CEL files and annotation files were downloaded, and the gene expression data of all samples were preprocessed via background correction, quantile normalization, and probe summarization using the Robust Multi-array Average (RMA) algorithm (http://www.bioconductor.org/packages/release/bioc/html/affy.html, accessed on 1 February 2022) in R 3.4.1 (R Studio, USA). Linear Models of Microarray Data package (LIMMA, version 3.32.5) from the link http://www.bioconductor.org/packages/release/bioc/html/limma.html (accessed on 1 February 2022) was used to identify DEGs [39].

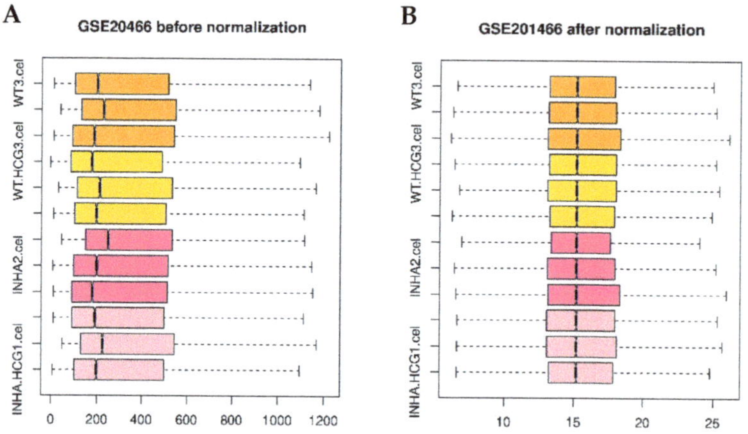

Figure 10. Boxplot of GSE20466 data preprocessing before (**A**) and after normalization (**B**). Pink, hot pink, gold, and orange boxes refer to Inha−/−(hCG), Inha−/−, WT(hCG), and WT samples.

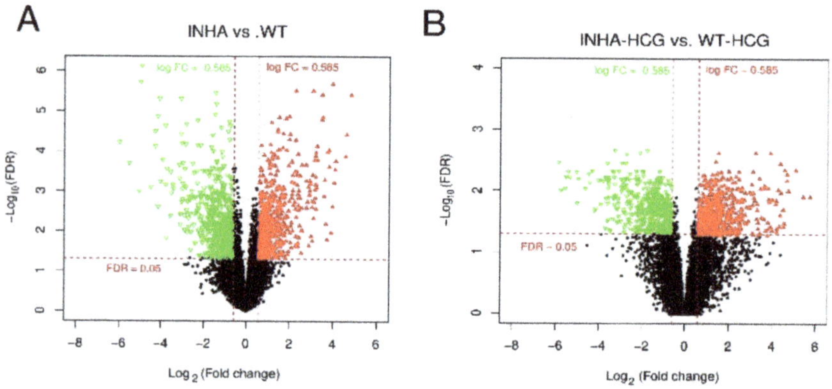

Figure 11. Volcano plot of DEGs in group Inha−/− vs. WT (**A**) and Inha−/− (hCG) vs. WT (hCG) (**B**). The red triangle and green inverted triangle refer to up and down-regulated DEGs; the Red horizontal dot line means FDR = 0.05 cutoff line, and two red vertical dot lines mean logFC = 1 and logFC = −1 cutoff line.

4.4. Hierarchical Clustering and Comparison Analysis of Selected DEGs in Different Groups

The expression of selected DEGs in Inha−/− vs. WT and Inha−/− (hCG) vs. WT (hCG), Inha−/− vs. Inha−/− (hCG), and WT vs. WT (hCG) were used to generate a hierarchical clustering image by heatmap (version 1.0.8) package in R 3.4.1 (RStudio; http://www.cran.r-project.org/web/packages/pheatmap/, accessed on 1 February 2022) [40,41]. Then, the identified DEGs were compared in the two groups and the ODEGs were selected by using VennDiagram package R 3.4.1 (http://www.cran.r-project.org/web/packages/VennDiagram/, accessed on 1 February 2022). After that, Pearson Correlation Coefficient (PCC) was used for further studying the ODEGs correlations following Huang da, et al. [42].

4.5. Enrichment Analysis for the Overlapping DEGs

To explore the functions of ODEGs and their pathways, the DAVID version 6.8 (Database for Annotation, Visualization and Integrated Discovery; http://www.david.ncifcrf.gov/, accessed on 1 February 2022) database was used to perform GO (Go Ontology) and KEGG (Kyoto Encyclopedia of Genes and Genomes) pathway enrichment analyses for

ODEGs. The p-value < 0.05 and gene count ≥ 2 were set as the cut-off criteria. Furthermore, the category of enriched GO, KEGG terms, and the gene number were displayed as scatterplots by the ggplot2 package in R3.4.1 (http://www.cran.r-project.org/web/packages/ggplot2/, accessed on 1 February 2022).

4.6. Physiological Phenotypes-Related Modules and Genes Identification Based on WGCNA

Weighted Gene Co-expression Network Analysis (WGCNA) algorithm was used to investigate the co-expression modules and genes which were related to phenotypes through the WGCNA package (version 1.61) (http://www.cran.r-project.org/web/packages/WGCNA/index.html, accessed on 1 February 2022).

4.7. Co-Expression Network Construction

Based on the results of the WGCNA algorithm, only gene pairs with expression correlation coefficient > 0.8 were used to construct a gene co-expression network which was then built by Cytoscape3.2.0 (http://www.cytoscape.org/, accessed on 1 February 2022). Also, GO and KEGG pathway enrichment analysis for the genes in the co-expression network was made.

4.8. miRNA-DEGs-TF Target Regulatory Network Analysis

WEB-based Gene Set Analysis Toolkit (WebGestalt; http://www.webgestalt.org/option.php, accessed on 1 February 2022) was used to search Transcription Factors (TFs) and miRNAs that regulated the DEGs in co-expression networks. Besides, p-value < 0.05 was set as the significance cut-off criteria. As a result of TFs and miRNAs were integrated and then identified in the miRNA-DEGs-TF regulatory network. The regulatory network consisting of DEGs, miRNAs, and TFs was then constructed and visualized by Cytoscape3.2.0 (http://www.cytoscape.org/, accessed on 1 February 2022).

4.9. Principal Component Analysis (PCA) for Genes in the Regulatory Network

In order to refine genes and get the most specific ones, we further narrowed the gene range by using the PCA algorithm in the psych package (version 1.7.5) in R3.1.4 (http://www.cran.r-project.org/web/packages/psych/, accessed on 1 February 2022). Then scatterplot3d package (version 0.3-40) (http://www.cran.r-project.org/web/packages/scatterplot3d/, accessed on 1 February 2022) was used to display the effect of PCA based on the top 3 components: PC1, PC2, and PC3.

5. Conclusions

In this study, 1074 and 931 DEGs aggregates were identified in inha and wild-type. Through bioinformatics investigation 8 miRNAs, 7 TFs and 14 DEGs and 7 up-regulated genes with 20 miRNA-DEGs regulation and 36 TFs-DEGs were confirmed. This study provides potential key information for using ODEGs as biomarkers for granulosa cell cancer regulation. Also, further integration of the DEGs and the TF related to the immune response can facilitate the development of the target drugs for controlling the transcription pathways of the inhibin-deficient females.

Supplementary Materials: The following supporting information can be downloaded at: https://www.mdpi.com/article/10.3390/molecules27175595/s1.

Author Contributions: Conceptualization, H.S.T., L.Y., M.G. and Z.u.R.; methodology, F.F. and M.S.H.; software and M.S.H.; formal analysis, A.L., M.G. and M.S.H.; validation, H.S.T., Z.u.R., M.G., F.F. and M.S.H.; investigation, F.F., H.S.T., M.G. and M.S.H.; resources, F.F., I.B. and M.S.H.; data curation, F.F., H.S.T., M.G., I.B. and M.S.H.; writing—original draft preparation, F.F., M.G., Z.u.R. and writing—review and editing, M.G., L.Y., F.F. and M.S.H.; visualization, M.G., F.F., I.B. and M.S.H.; supervision, L.Y. All authors have read and agreed to the published version of the manuscript.

Funding: This research and its APC were funded by Earmarked Fund for Modern Agro-industry Technology Research System (No. CARS-37-04B) and the National Natural Science Foundation of China (31272446).

Informed Consent Statement: Not applicable.

Data Availability Statement: Available upon request.

Conflicts of Interest: The authors declare no conflict of interest.

References

1. Dufour, S.; Quérat, B.; Tostivint, H.; Pasqualini, C.; Vaudry, H.; Rousseau, K. Origin and Evolution of the Neuroendocrine Control of Reproduction in Vertebrates, With Special Focus on Genome and Gene Duplications. *Physiol. Rev.* **2020**, *100*, 869–943 [CrossRef] [PubMed]
2. Filatov, M.; Khramova, Y.; Parshina, E.; Bagaeva, T.; Semenova, M. Influence of gonadotropins on ovarian follicle growth and development in vivo and in vitro. *Zygote* **2017**, *25*, 235–243. [CrossRef] [PubMed]
3. Kandaraki, E.A.; Chatzigeorgiou, A.; Papageorgiou, E.; Piperi, C.; Adamopoulos, C.; Papavassiliou, A.G.; Koutsilieris, M.; Diamanti-Kandarakis, E. Advanced glycation end products interfere in luteinizing hormone and follicle stimulating hormone signaling in human granulosa KGN cells. *Exp. Biol. Med.* **2018**, *243*, 29–33. [CrossRef] [PubMed]
4. Haas, J.; Bassil, R.; Meriano, J.; Samara, N.; Barzilay, E.; Gonen, N.; Casper, R.F. Does daily co-administration of letrozole and gonadotropins during ovarian stimulation improve IVF outcome? *Reprod. Biol. Endocrinol.* **2017**, *15*, 70. [CrossRef]
5. Han, L.; Wu, C.; Riaz, H.; Bai, L.; Chen, J.; Zhen, Y.; Guo, A.; Yang, L. Characterization of the mechanism of inhibin alpha-subunit gene in mouse anterior pituitary cells by RNA interference. *PLoS ONE* **2013**, *8*, e74596.
6. Calcaterra, V.; Cena, H.; Regalbuto, C.; Vinci, F.; Porri, D.; Verduci, E.; Chiara, M.; Zuccotti, G.V. The Role of Fetal, Infant, and Childhood Nutrition in the Timing of Sexual Maturation. *Nutrients* **2021**, *13*, 419. [CrossRef]
7. Robertson, D.M.; Pruysers, E.; Jobling, T. Inhibin as a diagnostic marker for ovarian cancer. *Cancer Lett.* **2007**, *249*, 14–17. [CrossRef]
8. Matzuk, M.M.; Finegold, M.J.; Su, J.G.; Hsueh, A.J.; Bradley, A. Alpha-inhibin is a tumour-suppressor gene with gonadal specificity in mice. *Nature* **1992**, *360*, 313–319. [CrossRef]
9. Rathore, R.; Arora, D.; Agarwal, S.; Sharma, S. Correlation of foxl2 with inhibin and calretinin in the diagnosis of ovarian sex cord stromal tumors. *Turk. J. Pathol.* **2017**, *33*, 121–128. [CrossRef]
10. Doroszko, M.; Chrusciel, M.; Belling, K.; Vuorenoja, S.; Dalgaard, M.; Leffers, H.; Nielsen, H.B.; Huhtaniemi, I.; Toppari, J.; Rahman, N.A. Novel genes involved in pathophysiology of gonadotropin-dependent adrenal tumors in mice. *Mol. Cell. Endocrinol.* **2017**, *444*, 9–18. [CrossRef]
11. Haverfield, J.T.; Stanton, P.G.; Loveland, K.L.; Zahid, H.; Nicholls, P.K.; Olcorn, J.S.; Makanji, Y.; Itman, C.M.; Simpson, E.R.; Meachem, S.J. Suppression of Sertoli cell tumour development during the first wave of spermatogenesis in inhibin α-deficient mice. *Reprod. Fertil. Dev.* **2017**, *29*, 609–620. [CrossRef]
12. Hetzler, K.L.; Hardee, J.P.; LaVoie, H.A.; Murphy, E.A.; Carson, J.A. Ovarian function's role during cancer cachexia progression in the female mouse. *Am. J. Physiol. Endocrinol. Metab.* **2017**, *312*, E447–E459. [CrossRef]
13. Nagaraja, A.K.; Agno, J.E.; Kumar, T.R.; Matzuk, M.M. Luteinizing hormone promotes gonadal tumorigenesis in inhibin-deficient mice. *Mol. Cell. Endocrinol.* **2008**, *294*, 19–28. [CrossRef]
14. Kumar, T.R.; Palapattu, G.; Wang, P.; Woodruff, T.K.; Boime, I.; Byrne, M.C.; Matzuk, M.M. Transgenic models to study gonadotropin function: The role of follicle-stimulating hormone in gonadal growth and tumorigenesis. *Mol. Endocrinol.* **1999**, *13*, 851–865. [CrossRef]
15. Chermuła, B.; Brązert, M.; Iżycki, D.; Ciesiółka, S.; Kranc, W.; Celichowski, P.; Ożegowska, K.; Nawrocki, M.J. New Gene Markers of Angiogenesis and Blood Vessels Development in Porcine Ovarian Granulosa Cells during Short-Term Primary Culture In Vitro. *BioMed Res. Intl.* **2019**, *2019*, 6545210. [CrossRef]
16. Li, Z.; Wang, J.; Zhao, Y. scRNA-seq of ovarian follicle granulosa cells from different fertility goats reveals distinct expression patterns. *Reprod. Domest. Anim.* **2021**, *56*, 801–811. [CrossRef]
17. Nia, A.M.; Chen, T.; Barnette, B.L.; Khanipov, K.; Ullrich, R.L.; Bhavnani, S.K.; Emmett, M.R. Efficient identification of multiple pathways: RNA-Seq analysis of livers from (56)Fe ion irradiated mice. *BMC Bioinform.* **2020**, *21*, 118. [CrossRef]
18. Langfelder, P.; Horvath, S. WGCNA: An R package for weighted correlation network analysis. *BMC Bioinform.* **2008**, *9*, 559. [CrossRef]
19. Gouda, M.; Huang, Z.; Liu, Y.; He, Y.; Li, X. Physicochemical impact of bioactive terpenes on the microalgae biomass structural characteristics. *Bioresour. Technol.* **2021**, *334*, 125232. [CrossRef]
20. Gouda, M.; Chen, K.; Li, X.; Liu, Y.; He, Y. Detection of microalgae single-cell antioxidant and electrochemical potentials by gold microelectrode and Raman micro-spectroscopy combined with chemometrics. *Sens. Actuators B Chem.* **2021**, *329*, 129229. [CrossRef]

21. Shankar, V.; Gouda, M.; Moncivaiz, J.; Gordon, A.; Reo, N.V.; Hussein, L.; Paliy, O. Differences in Gut Metabolites and Microbial Composition and Functions between Egyptian and U.S. Children Are Consistent with Their Diets. *mSystems* **2017**, *2*, e00169-16. [CrossRef]
22. FarmanUllah; Liang, X.; Khan, F.; Salim, M.; Rehman, Z.; Khan, M.; Talpur, H.; Schreurs, N.; Gouda, M.; Khan, S.; et al. Transcriptomic in silico analysis of bovine Escherichia coli mastitis highlights its immune-related expressed genes as an effective biomarker. *J. Genet. Eng. Biotechnol.* **2021**, *19*, 00290–00291. [CrossRef]
23. Ahmed, F.E.; Gouda, M.M.; Hussein, L.A.; Ahmed, N.C.; Vos, P.W.; Mohammad, M.A. Role of Melt Curve Analysis in Interpretation of Nutrigenomics' MicroRNA Expression Data. *Cancer Genom. Proteom.* **2017**, *14*, 469–481.
24. Fong, M.Y.; Kakar, S.S. Ovarian cancer mouse models: A summary of current models and their limitations. *J. Ovarian Res.* **2009**, *2*, 12. [CrossRef] [PubMed]
25. Kadariya, I.; Wang, J.; ur Rehman, Z.; Ali, H.; Riaz, H.; He, J.; Bhattarai, D.; Liu, J.J.; Zhang, S.J. RNAi-mediated knockdown of inhibin α subunit increased apoptosis in granulosa cells and decreased fertility in mice. *J. Steroid Biochem. Mol. Biol.* **2015**, *152*, 161–170. [CrossRef]
26. Vasilache, A.M.; Kugelberg, U.; Blomqvist, A.; Nilsberth, C. Minor changes in gene expression in the mouse preoptic hypothalamic region by inflammation-induced prostaglandin E2. *J. Neuroendocrinol.* **2013**, *25*, 635–643. [CrossRef]
27. Huang, W.; Zhang, X.; Li, A.; Xie, L.; Miao, X. Genome-Wide Analysis of mRNAs and lncRNAs of Intramuscular Fat Related to Lipid Metabolism in Two Pig Breeds. *Cell. Physiol. Biochem.* **2018**, *50*, 2406–2422. [CrossRef]
28. Chen, C.; Ahmad, M.J.; Ye, T.; Du, C.; Zhang, X.; Liang, A.; Yang, L. Cathepsin B Regulates Mice Granulosa Cells' Apoptosis and Proliferation In Vitro. *Int. J. Mol. Sci.* **2021**, *22*, 11827. [CrossRef]
29. Jing, R.; Gu, L.; Li, J.; Gong, Y. A transcriptomic comparison of theca and granulosa cells in chicken and cattle follicles reveals ESR2 as a potential regulator of CYP19A1 expression in the theca cells of chicken follicles. *Comp. Biochem. Physiol. Part D Genom. Proteom.* **2018**, *27*, 40–53. [CrossRef]
30. Chen, Y.; Yang, L.; Qin, Y.; Liu, S.; Qiao, Y.; Wan, X.; Zeng, H.; Tang, X.; Liu, M.; Hou, Y. Effects of differential distributed-JUP on the malignancy of gastric cancer. *J. Adv. Res.* **2021**, *28*, 195–208. [CrossRef]
31. Peng, M.; Zhang, H.; Jaafar, L.; Risinger, J.I.; Huang, S.; Mivechi, N.F.; Ko, L. Human ovarian cancer stroma contains luteinized theca cells harboring tumor suppressor gene GT198 mutations. *J. Biol. Chem.* **2013**, *288*, 33387–33397. [CrossRef] [PubMed]
32. Leask, A.; Abraham, D.J. All in the CCN family: Essential matricellular signaling modulators emerge from the bunker. *J. Cell Sci.* **2006**, *119*, 4803–4810. [CrossRef] [PubMed]
33. Fang, J.; Xiao, L. Junction plakoglobin, a potential prognostic marker of oral squamous cell carcinoma, promotes proliferation, migration and invasion. *J. Oral Pathol. Med.* **2020**, *49*, 30–38. [CrossRef] [PubMed]
34. Zangen, D.; Kaufman, Y.; Zeligson, S.; Perlberg, S.; Fridman, H.; Kanaan, M.; Abdulhadi-Atwan, M.; Abu Libdeh, A.; Gussow, A.; Kisslov, I.; et al. XX ovarian dysgenesis is caused by a PSMC3IP/HOP2 mutation that abolishes coactivation of estrogen-driven transcription. *Am. J. Hum. Genet.* **2011**, *89*, 572–579. [CrossRef]
35. Yin, T.; Getsios, S.; Caldelari, R.; Kowalczyk, A.P.; Müller, E.J.; Jones, J.C.R.; Green, K.J. Plakoglobin suppresses keratinocyte motility through both cell–cell adhesion-dependent and -independent mechanisms. *Proc. Natl. Acad. Sci. USA* **2005**, *102*, 5420–5425. [CrossRef]
36. Hofland, J.; Steenbergen, J.; Voorsluijs, J.M.; Verbiest, M.M.; de Krijger, R.R.; Hofland, L.J.; de Herder, W.W.; Uitterlinden, A.G.; Feelders, R.A.; de Jong, F.H. Inhibin alpha-subunit (INHA) expression in adrenocortical cancer is linked to genetic and epigenetic INHA promoter variation. *PLoS ONE* **2014**, *9*, e104944. [CrossRef]
37. Nagaraja, A.K.; Middlebrook, B.S.; Rajanahally, S.; Myers, M.; Li, Q.; Matzuk, M.M.; Pangas, S.A. Defective gonadotropin-dependent ovarian folliculogenesis and granulosa cell gene expression in inhibin-deficient mice. *Endocrinology* **2010**, *151*, 4994–5006. [CrossRef]
38. Chen, L.; Zhang, W.; Huang, R.; Miao, X.; Li, J.; Yu, D.; Li, Y.; Hsu, W.; Qiu, M.; Zhang, Z.; et al. The function of Wls in ovarian development. *Mol. Cell. Endocrinol.* **2021**, *522*, 111142. [CrossRef]
39. Wang, L.; Cao, C.; Ma, Q.; Zeng, Q.; Wang, H.; Cheng, Z.; Zhu, G.; Qi, J.; Ma, H.; Nian, H.; et al. RNA-seq analyses of multiple meristems of soybean: Novel and alternative transcripts, evolutionary and functional implications. *BMC Plant Biol.* **2014**, *14*, 169. [CrossRef]
40. Chen, H.; Boutros, P.C. VennDiagram: A package for the generation of highly-customizable Venn and Euler diagrams in R. *BMC Bioinform.* **2011**, *12*, 35. [CrossRef]
41. Diao, C.; Xi, Y.; Xiao, T. Identification and analysis of key genes in osteosarcoma using bioinformatics. *Oncol. Lett.* **2018**, *15*, 2789–2794. [CrossRef]
42. Huang da, W.; Sherman, B.T.; Lempicki, R.A. Systematic and integrative analysis of large gene lists using DAVID bioinformatics resources. *Nat. Protoc.* **2009**, *4*, 44–57. [CrossRef]

molecules

MDPI

Article

The Effect of Combining Post-Harvest Calcium Nanoparticles with a Salicylic Acid Treatment on Cucumber Tissue Breakdown via Enzyme Activity during Shelf Life

Mohamed F. M. Abdelkader [1], Mohamed H. Mahmoud [2], Lo'ay A. A. [3,*], Mohamed A. Abdein [4,*], Khaled Metwally [5,6], Shinya Ikeno [6,*] and Samar M. A. Doklega [7]

1 Department of Plant Production, College of Food and Agriculture, King Saud University, Riyadh 12372, Saudi Arabia; mohabdelkader@ksu.edu.sa
2 Department of Biochemistry, College of Science, King Saud University, Riyadh 12372, Saudi Arabia; mmahmoud2@ksu.edu.sa
3 Pomology Department, Faculty of Agriculture, Mansoura University, Mansoura 35516, Egypt
4 Seed Development Department, Agricultural Professions Syndicate, Downtown, Cairo 11669, Egypt
5 Department of Genetics, Faculty of Agriculture, Ain Shams University, Cairo 11241, Egypt; khaleda.fatah@agr.asu.edu.eg or metwally.khaled-abdelfatteh735@mail.kyutech.jp
6 Department of Biological Functions Engineering, Graduate School of Life Science and Systems Engineering, Kyushu Institute of Technology, 2-4 Hibikino, Wakamatsu, Kitakyushu 808-0196, Japan
7 Floriculture and vegetable Department, Faculty of Agriculture, Mansoura University, Mansoura 35516, Egypt; samar_2005@mans.edu.eg
* Correspondence: loay_arafat@mans.edu.eg (L.A.A.); abdeingene@yahoo.com (M.A.A.); ikeno@life.kyutech.ac.jp (S.I.)

Abstract: In the present study, an experiment was carried out on the postharvest of cucumber fruit during a 14-day shelf life. The aim was to assess the impact of calcium nanoparticles (CaNPs) blended with different concentrations of salicylic acid (SA) on the shelf life of cucumbers during the seasons of 2018 and 2019. The investigation further monitored the influences of CaNPs-SA on some physical properties of cucumber, including the percentage weight loss, color, and fruit firmness. In addition, chemical properties, such as total soluble solids (SSC%), total acidity (TA%), total soluble sugars, and chlorophyll pigmentation of the fruit skin, were assessed during a 14-day shelf lifeCell wall degradation enzymes (CWEAs) such as polygalacturonase (PG), cel-lulase (CEL), xylanase (XYL), and pectinase (PT) were also researched. In addition, the generation rates of H_2O_2 and $O_2^{\bullet -}$ were calculated, as well as the reduction of DPPH. The lipid peroxidation (malondialdehyde, MDA) and cell membrane permeability (IL%) of cell wall composites were also determined. CaNPs-SA at 2 mM suppressed CWEAs, preserved fruit quality, reduced weight loss throughout the shelf-life period, and reduced the percent leakage value. At this concentration, we also found the lowest levels of MDA and the highest levels of DPPH.

Keywords: shelf life; cucumber; cell wall degradation enzymes; nano calcium particles and salicylic acid

1. Introduction

In Egypt, cucumber (*Cucumber sativa* L. cv Barracuda) is the most popular vegetable crop grown for local and global consumption. Additionally, it is a preferred product around the world Cucumber growth generally increases throughout the summer season; in Egypt, a total area of 226,385 ha is used to grow cucumber, yielding approximately 364,571 tons [1]. Cucumber quality decreases significantly after harvesting due to a loss of water, shriveling, and yellowing from loosening skin chlorophyll pigment; as a result, cucumbers have a shorter shelf life in the market, lasting about 2–3 days [2]. Cucumber is a non-climacteric vegetable crop that contains more than 90% water. The primary problem that influences fruit quality during processing is often excessive moisture loss [3]. The outcomes of weight loss result in a lower marketable income. In addition, fruits are more sensitive to infection by postharvest pathogens [4]. Naturally, fruits and vegetables have many layers of wax on their surfaces [5]. The wax diminishes the rate of water evaporation in fruit tissues during the growth, development, or post-harvest stages [6]. These layers vary in thickness throughout the circulation stages and are affected by processing stages, such as washing, where they are easy to remove [7]. Moreover, many variables promote the deterioration and waste of fruits during storage, including environmental factors [8], harvest time, the stage of fruit maturity, and the occurrence of mechanical damage throughout handling phases [9]. Previously, coating techniques have been implemented to improve the shelf life and quality of fruits and vegetables [10], or to develop deliberate cucumber characteristics throughout the selling chain for consumers [11]. However, deciding on the right coating blend assuredly affects the performance of other coating layers. Choosing a film-type coating for fruit is typically a safe and effective way to preserve cucumber quality by reducing fruit rotting and minimizing weight loss. [12].

The recent application of advanced techniques, such as nanotechnology, in the post-harvest of fruits and vegetables merits further investigation. The nano-technique, which works to improve a material's physical and chemical properties, also has significant antifungal and antiviral properties. [13]. It also has uses in different fields, such as medicine [14] and pharmacology [15]. Furthermore, in agriculture for horticultural products, nanotechnology has a positive effect on the preservation of fruits and vegetables during storage, thereby increasing their shelf life [16].

Salicylic acid (SA) is recognized as a phenolic compound that has effects on plant growth and defenses against different stresses that occur simultaneously during the postharvest of fruit [17]. Fruit tissues (cells) produce reactive oxygen species (ROS) when exposed to stress conditions [18]. ROS include superoxide ($^1O_2$), peroxide (H_2O_2), and hydroxyl ($OH^\bullet$) ions, which cause damage under stress conditions [19]. As a result of stress damage in cellular structures, ROS are also produced. [20]. In addition, SA controls multiple physiological and biochemical paths in cells [21]. It inhibits the harmful effects of these ROS by enhancing antioxidant activities, such as lowering H2O2 levels through the action of ascorbate (APX) [22]. Additionally, SA restrains fruit senescence throughout the duration of storage [23,24], maintains elements of fruit quality [25], reduces fungal rotting infection in susceptible vegetables [13,26], and efficiently influences resistance to chilling [27].

Therefore, in this investigation, cucumber 'Barracuda' fruits were treated with CaNPs blended with salicylic acid at varying doses and then stored at the ambient condition. Enzymatic activity was then measured to see how the CaNPs-SA blends affected tissue breakdown over the course of the shelf-life.

2. Results

2.1. Physical Properties

Figure 1 illustrates the variations in the physical properties of cucumber fruit (weight loss percentage, fruit color hue angle, and firmness) during shelf life. There was a significant impact of CaNPs-SA applications on cucumber fruits at $p < 0.05$ when considering storage duration (days) as a factor. It can be seen from the figure that the variations between CaNPs-SA treatments for the weight loss percentage of cucumber fruits were remarkable on the fourth day of shelf life. The weight loss percentage increased gradually overall among all CaNPs-SA treatments during shelf durations. Cucumber fruits treated with 2 mmol L^{-1} CaNPs-SA had a significantly lower weight loss percentage (19.74%) on the 14th day of storage. However, at the end of the experiment, other CaNPs-SA treatments independently showed a higher weight loss percentage at the same recorded interval—control (37.35 percent), CaNPs-SA 0 mM (35.14 percent), and CaNPs-SA 1 mM (30.25 percent). The fruit color hue angle (h°) was also monitored during the 14-day shelf life. Cucumber color (h°) decreased independently according to CaNPs-SA treatments during 14 days of storage. The fruits treated with CaNPs-SA in 2 mmol L^{-1} treatments showed a high hue angle throughout the storage duration. In addition, cucumber fruit firmness (Figure 1) showed an interaction between CaNPs-SA treatment and shelf-life duration (days) at $p < 0.001$ when both were admitted as experimental factors. Fruit firmness was higher at harvest time and decreased continually and gradually up until the end of the shelf-life period. It was noted that the treatment of cucumber fruits with CaNPs-SA at 2 mmol L^{-1} better maintained fruit firmness compared to the other treatments during the storage period.

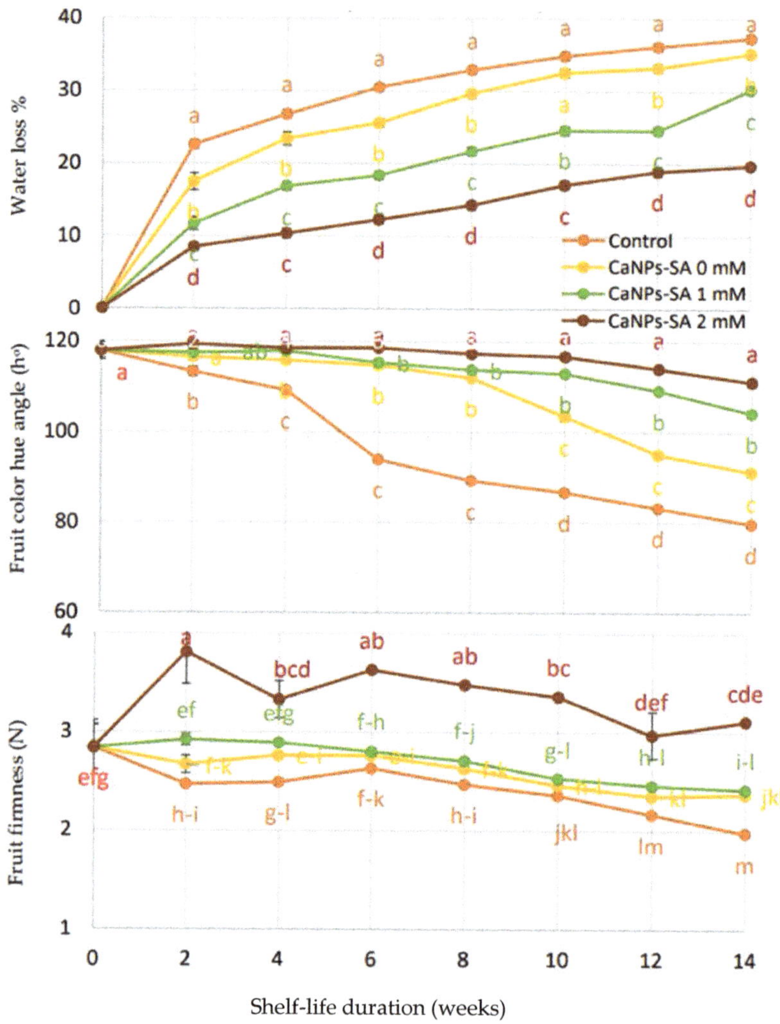

Figure 1. The effects of CaNPs blended with salicylic acid at various doses (0, 1, and 2 mmol L^{-1}) on cucumber physical attributes (weight loss percent, fruit color, and firmness) over time. The standard error ($n = 3$) is represented by the vertical bar for the mean of the two seasons.

2.2. SSC%, TA%, and SSC/TA Ratio

Figure 2 shows a significant interaction at $p \leq 0.001$ between shelf-life duration in days and CaNPs-SA application. In comparison to the initial value at harvest time, the total soluble solid content (SSC percent) grew modestly and steadily throughout the CaNPs-SA treatments up to the end of the shelf-life period. However, TA percent declined during storage, with larger declines than the original levels at harvest. When compared to control fruits and fruits treated with CaNPs-SA during shelf life, the coated CaNPs-SA treatment at 2 mM provided greater stability in these parameters. On the 14th day of shelf life, the SSC percentage was 2.73 percent, the TA percentage was 0.062 percent, and the total sugar percentage was 0.54 percent, compared to the control fruits, which had an SSC percentage of 3.92 percent, a TA percentage of 0.0.039 percent, and a total sugar percentage of 0.62 percent.

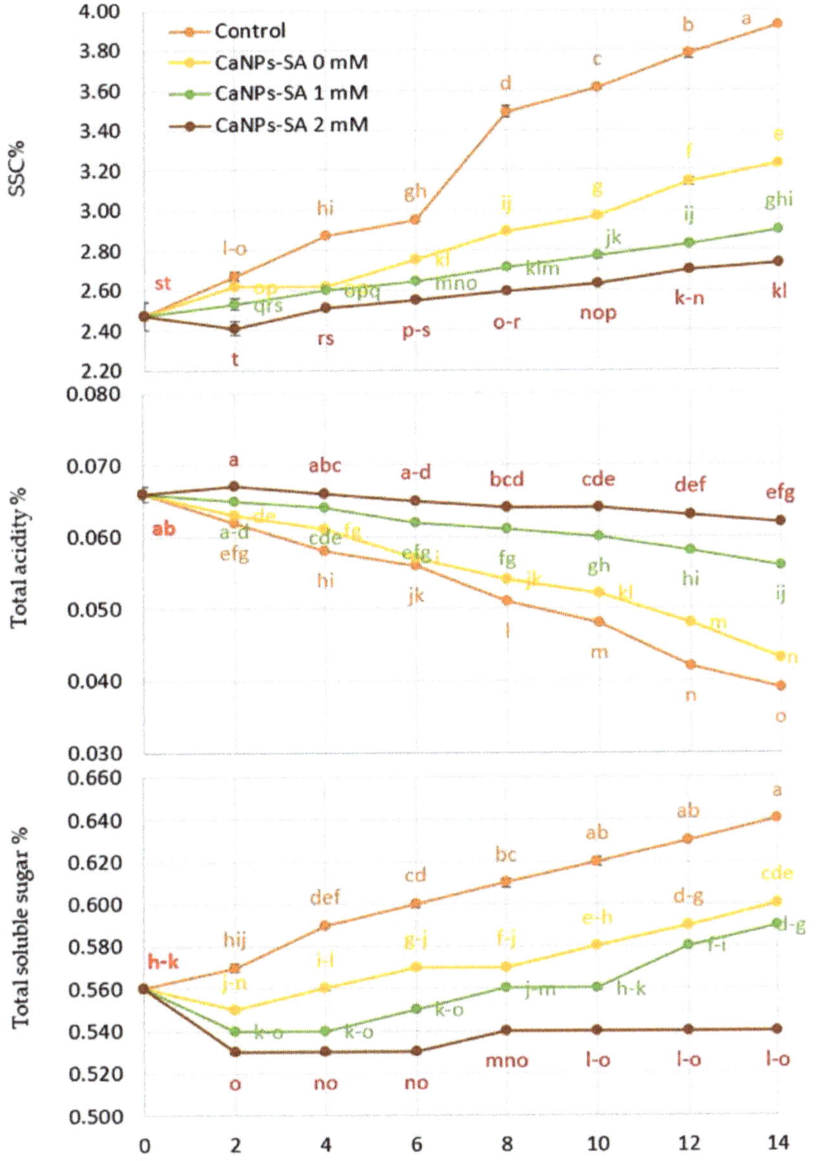

Figure 2. The effects of CaNPs blended with salicylic acid at various doses (0, 1, and 2 mmol L^{-1}) on cucumber chemical attributes (SSC%, total acidity %, and total soluble sugar %) over time. The standard error (n = 3) is represented by the vertical bar for the mean of the two seasons.

2.3. Cucumber Chlorophyll Pigment Content

Figure 3 shows the variations in fruit skin pigment (chlorophyll content) as a function of shelf-life duration in days. It can be seen from the figure that chlorophyll pigments presented a significant interaction between variables at $p \leq 0.001$ when the shelf-life duration (days) and CaNPs-SA applications were analyzed. Observably, all chlorophyll elements progressively decreased throughout the storage phase. The decreases were caused by a decrease in the salicylic acid content of the CaNPs. The fruits that were immersed in 2 mM CaNPs-SA treatments presented with greater preservation of chlorophyll compartments up to the end of the experiment period. The results were recorded (Chl_a = 6.79, Chl_b = 3.04, and Chl_{a+b} = 9.83 mg g^{-1} FW) compared to the initial value at harvest time and other treatments. The control cucumber treatment showed a more rapid degradation in chlorophyll pigment at the same interval (Chl_a = 5.84, Chl_b = 2.54, and Chl_{a+b} = 8.18 mg g^{-1} FW).

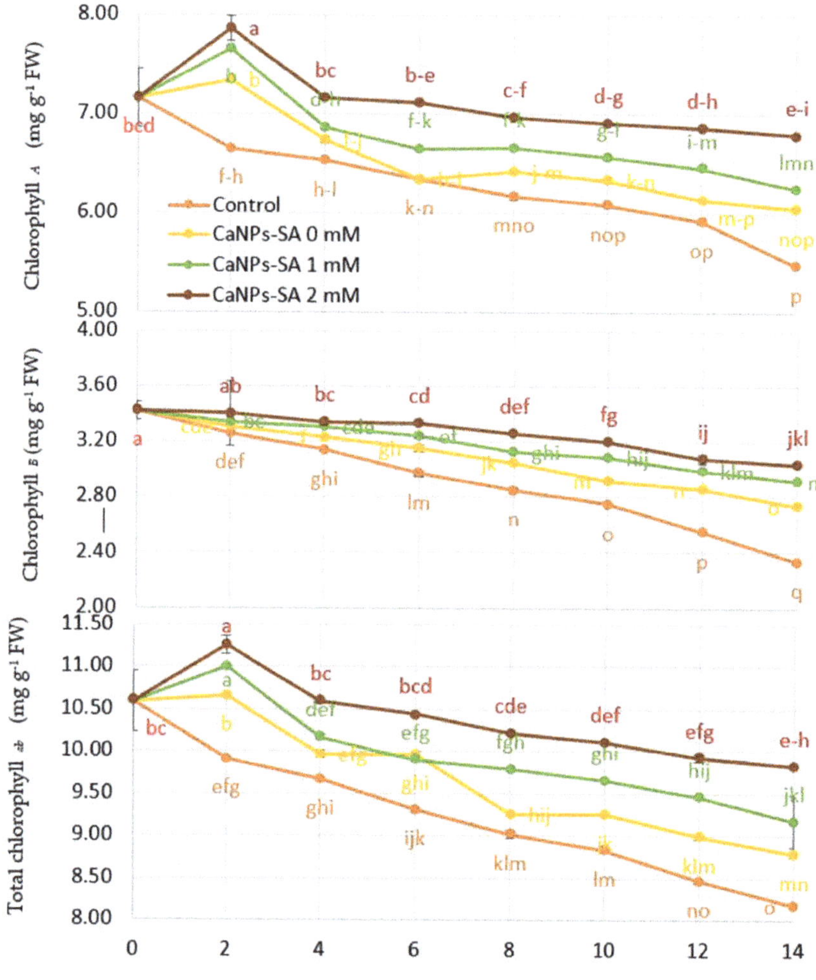

Figure 3. The effects of CaNPs blended with salicylic acid at various doses (0, 1, and 2 mmol L^{-1}) on cucumber chemical attributes (chlorophyll compartments) over time. The standard error (n = 3) is represented by the vertical bar for the mean of the two seasons.

2.4. Effect of CaNPs-SA Treatments on Cell Wall Degradation Enzyme Activities

Figure 4 shows the activities of cell wall-degrading enzymes (CWEAs), such as PG, CEL, XYL, and PT, during the shelf-life period (days) and CaNPs-SA treatments as factors. When time and CaNPs-SA treatments were considered, there was a significant interaction ($p < 0.001$). Initially, CWEAs increased gradually after harvesting and immersion in different CaNPs-SA treatments and continued to increase up to the 14th day of shelf life. The CaNPs-SA 2 mM treatment reduced overall CWEAs compared to the other treatments and control fruit. However, CWEAs increased more quickly in control fruits, and all enzymes increased until the 14th day of shelf life. Notably, the degrading enzyme activities increased independently of the CaNPs-SA treatments during shelf life and decreased with increasing SA concentrations. However, cucumber fruits that were treated with CaNPs-SA at 2 mM presented lower CWEAs during the 14-day shelf-life period.

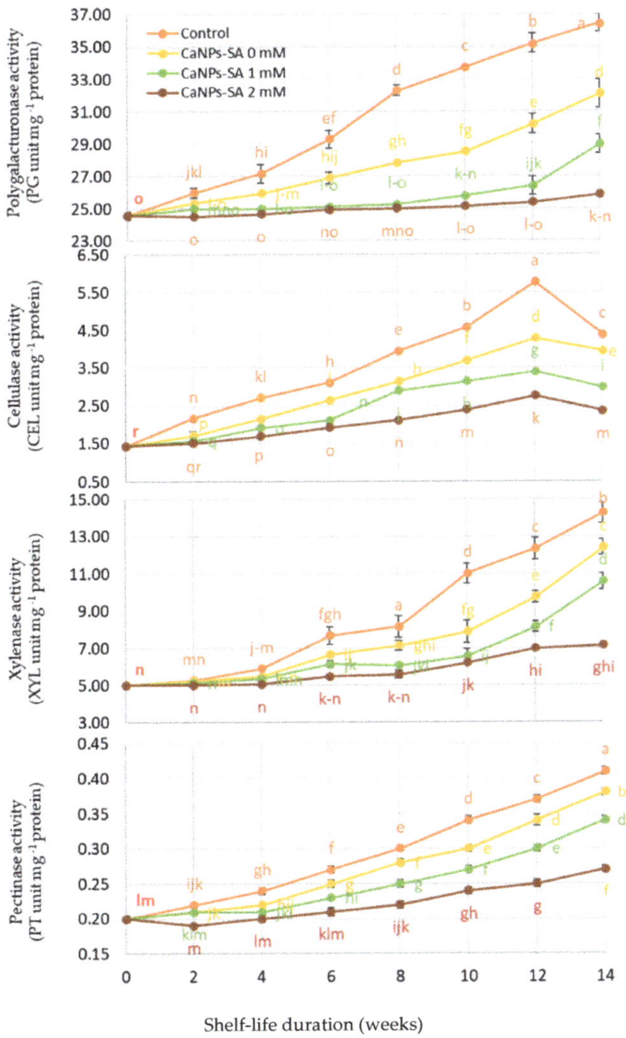

Figure 4. The effects of CaNPs blended with salicylic acid at various doses (0, 1, and 2 mmol L^{-1}) on cucumber chemical attributes (cell wall enzyme activities) over time. The standard error ($n = 3$) is represented by the vertical bar for the mean of the two seasons.

2.5. $O_2^{\bullet-}$ and H_2O_2 Production, DPPH Reduction, LOX, and Percent of Ion Leakage

The changes in $O_2^{\bullet-}$ and H_2O_2 output rate and the decrease in DPPH as a function of shelf-life duration are shown in Figures 5 and 6. When the shelf life (days) and CaNPs-SA treatments were evaluated, the $O_2^{\bullet-}$ and H_2O_2 formation rates showed a significant interaction at $p < 0.001$. During the 14-day shelf-life trial, the output of $O_2\bullet$ and H_2O_2 increased steadily for all applications. Over the course of the storage time, the control fruits produced more $O_2^{\bullet-}$ and H_2O_2 than the other treatments. On the 14th day of storage, however, fruits treated with CaNPs-SA at 2 mM exhibited considerably lower $O_2^{\bullet-}$ and H_2O_2 generation (0.32 and 0.10 nmol min^{-1} g^{-1} FW, respectively) compared to the control fruits (0.42 and 0.18 nmol min^{-1} g^{-1} FW, respectively).

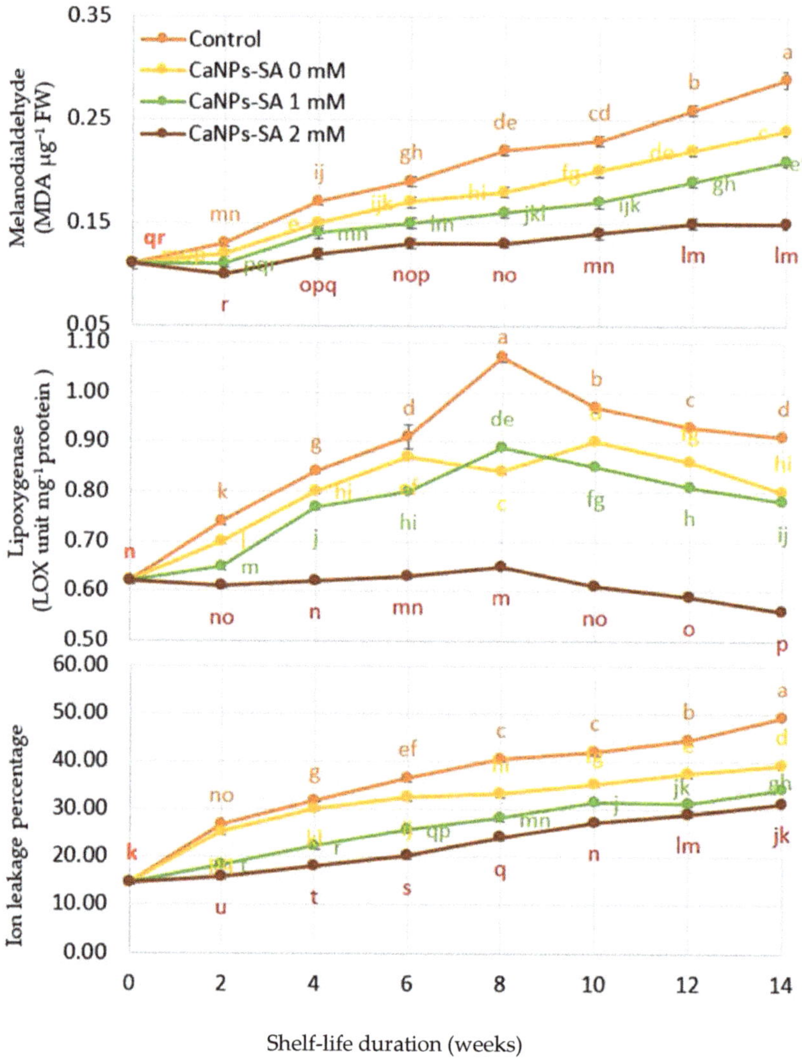

Figure 5. The effects of CaNPs blended with salicylic acid at various doses (0, 1, and 2 mmol L^{-1}) on cucumber chemical attributes (MDA, LOX, and ion leakage) over time. The standard error ($n = 3$) is represented by the vertical bar for the mean of the two seasons.

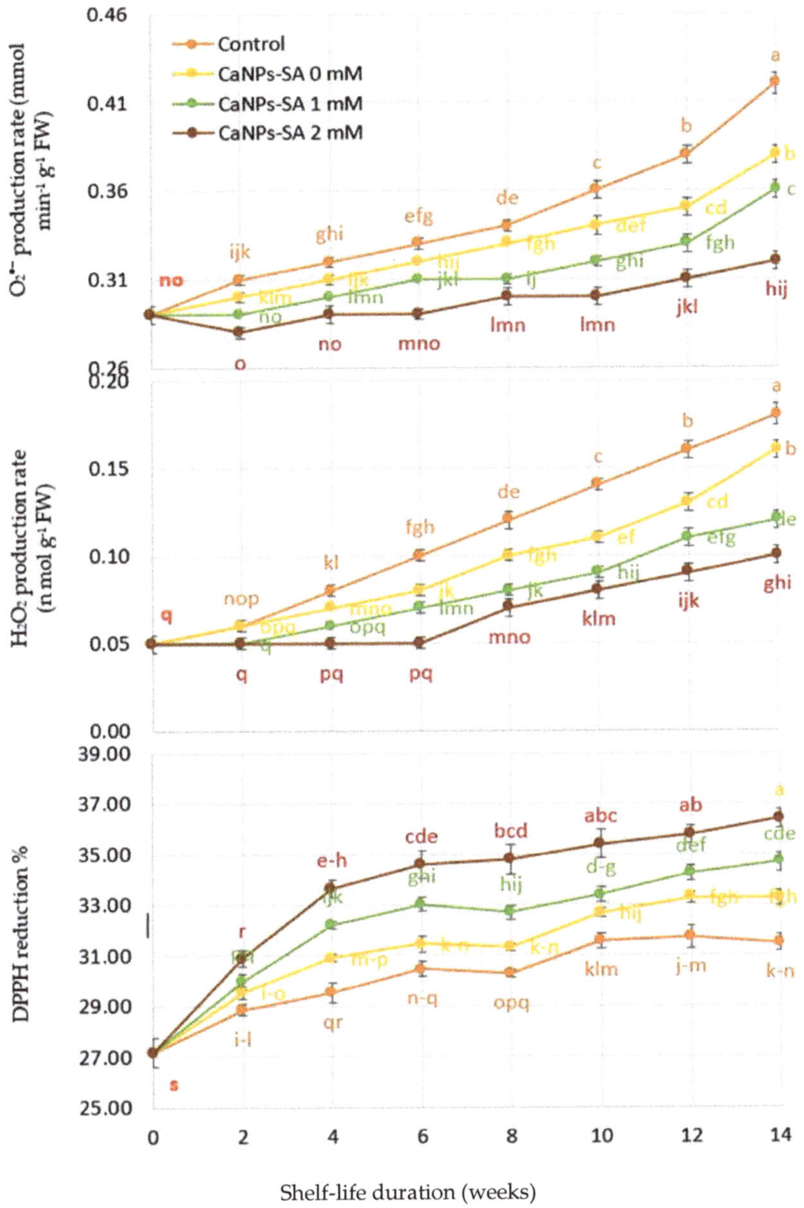

Figure 6. The effects of CaNPs blended with salicylic acid at various doses (0, 1, and 2 mmol L^{-1}) on cucumber chemical attributes ($O_2^{\bullet-}$, H_2O_2, and DPPH reduction rate) over time. The standard error (n = 3) is represented by the vertical bar for the mean of the two seasons.

3. Discussion

The observed variations in our results may be explained by the high concentration of SA blended with the calcium nanoparticles, which could have inhibited the cell wall degradation enzyme activities (CWEAs) during shelf life [28]. Our results confirmed that the CANPs-SA treatment at 2 mmol L^{-1} inhibited CWEAs considerably more than the other treatments and control fruit throughout storage period (Figure 4). As a result, the fruit firmness was affected. Fruits treated with CaNPs-SA at 2 mM were firmer during the shelf-life period (Figure 2). This could also be due to the increased inhibition of cell wall degradation enzymes, such as PG, CEL, and XYL, all of which are linked to fruit firmness [29]. Because of the presence of salicylic acid, the water content of the cucumber fruit tissue was preserved during the shelf life by inhibiting CWEAs [5,27,30,31].

These outcomes could well be correlated with the impact of treatment with CaNPs loaded with 2 mM salicylic acid. This treatment lowers water transpiration [32], delays fruit ripening/senescence [23], and activates antioxidant enzymes during storage, giving additional protection to the cucumber tissue/cells against oxidative reactions [33]. Antioxidant enzyme system activities are increased when CS/PVP and SA are combined at 2 mM, resulting in decreased lipid peroxidation accumulation [16,34–36].

The changes in chlorophyll compartments over time could be attributed to a decrease in antioxidant enzyme activity [37,38]. As a result, more ROS are created, resulting in increased oxidative reactions, such as lipid peroxidation and protein oxidation. Following this, further malfunction develops in cell membrane structures along with an increase in IL percent [39], finally leading to cell death [27,40,41].

As a result, the observed treatment effects were most likely the result of CaNPs and SA limiting cell wall hydrolysis by modulating CEL, PG, and PT. [23]. An improvement in fruit firmness throughout the shelf life was associated with a faster increase in endogenous SA in fruit tissue (Figure 1) [42]. Interestingly, the low level of CWEAs suggests that PG can inhibit SA during storage (Huber, 1983), and the reduction in CEL activity caused by CaNPs-SA at 2 mM could be linked to a structural change in the hemicellulose required for fruit firmness or softening [29]. According to the findings of [43], CaNPs in the presence of SA at 2 mM reduce fruit senescence or ripening by limiting CWEAs and suppressing ethylene production.

Reduced antioxidant enzyme activity could be the cause of the increases in MDA, LOX, and IL percent with control treatments over the shelf-life period [37]. As a result, more ROS are produced, resulting in increased oxidative responses in terms of lipid peroxidation and LOX activation. Following this, there is an increased disruption of the cell membrane structure and higher IL percent [16], followed by cell death [33].

Other metabolic activities, such as aerobic respiration, could be responsible for the high quantities of $O_2^{\bullet}$ and H_2O_2 produced during the experiment [44]. As a result, increasing AEAs, such as SOD, may improve fruit tissue tolerance to $O_2^{\bullet-}$, and increasing CAT and APX activities could aid in the scavenging of both $O_2^{\bullet-}$ and H_2O_2 during storage [45].

4. Materials and Methods

4.1. Fruit Materials

Immature cucumber (*Cucumis sativus* L. cv Barracuda) fruits were picked in 2018 and 2019 from a commercial farm in Mansoura, Dakahlia, Egypt. Upon arrival at the Mansoura University Floriculture and Vegetable Department, the fresh fruits were handpicked for uniformity in size and shape and the absence of external injury. The fruits were cleaned with chlorinated water (0.05% NaOCl), flushed with distilled water, and permitted to air dry. A total of 192 fruits were selected for the shelf-life experiment and were allocated toward pair lots; each lot contained 96 fruits. The first lot was used to measure the physical naturalistic variables and the second batch was designated for chemical analysis throughout the 14-day duration of shelf life. The physical and chemical examinations were judged every two days throughout the continuation of the research experiment.

4.2. Use of Salicylic Acid in Calcium Carbonate Nanoparticle (CaNP) Production

Calcium nanoparticles (CaNPs) were produced using a modified version of the procedure described in [46]. The CaNPs were designed by introducing SA at 1 and 2 mM with filtered water. The SA was introduced using pure water and $CaCl_2$ liquid at 50 mM. The mixture was shaken for one hour at 5000 rpm and then stored at room temperature for three days. The mixture was then shaken for another hour at 5000 rpm before being left at room temperature for three days.

4.3. Nanoparticle Characterization Utilizing UV-Vis Spectroscopy

Using the ATI Unicom UV-vis spectroscopic analysis image program (Ver. 3.20), the reduction in absolute Ca^{++} particles and the increase in succeeding calcium nanoparticles were detected by identifying the UV-Vis color curve of the response mixture at multiple wavelengths. The UV-Vis spectra of the mixed CaNPs were observed to be in the 240–440 nm range. This experiment was carried out at 25 °C using quartz cuvettes (1 cm optical way, Figure 7).

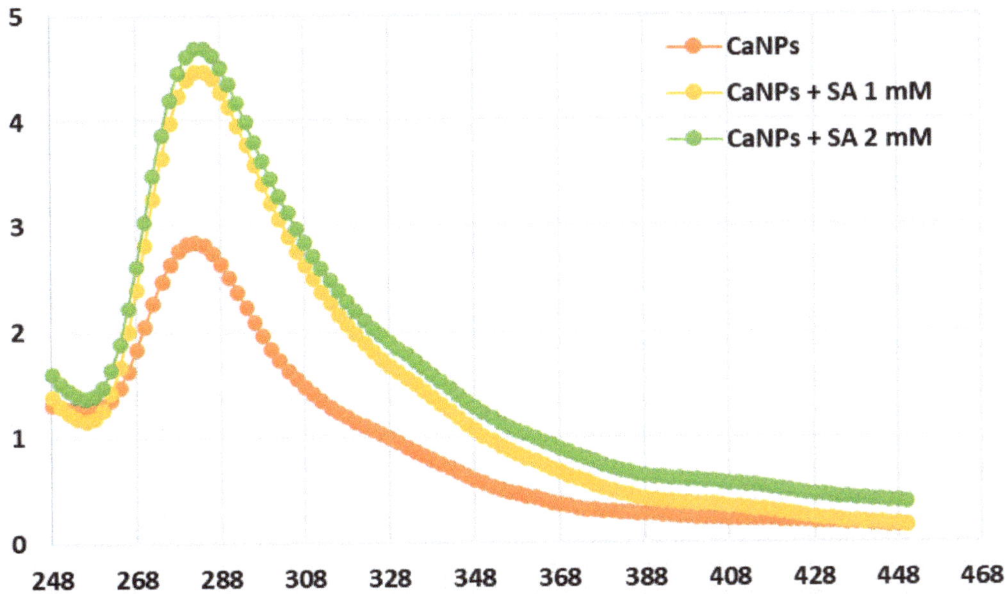

Figure 7. UV-visible absorption spectra of calcium nanoparticles synthesized (CaNPs) blended with various concentrations of salicylic acid (0, 1, and 2 mM) with a peak at 270 nm.

4.4. CaNP Characterization via Zeta Potential

Zeta potential analysis (ZPA) was used to define the situation of the CaNPs at the start and to check the equilibrium of the CaNP liquid at the end. The Electron Microscope Lab determined the combination of CaNPs, including the SA external charge, using the Malvern Instruments Ltd. Zeta Potential software (Ver. 2.3). The CaNPs-SA blend had an exterior charge that attracted a small layer of differently charged anions to its surface. CaNPs were encased in two layers of anions that moved around as they were dispersed throughout the mixture. The electric potential near the end of the duplicated layer is called the zeta potential, and it varies from +100 mV to −100 mV (Figure 8). After CaNPs were integrated with SA, the ZPA rate was 4.74 mV (highly stable). CaNPs with ZPA rates greater than or equal to +25 mV are usually considered as being more stable [47].

			Mean (mV)	Area (%)	St Dev (mV)
Zeta Potential (mV):	-2.86	Peak 1:	-2.86	100.0	6.09
Zeta Deviation (mV):	6.09	Peak 2:	0.00	0.0	0.00
Conductivity (mS/cm):	3.23	Peak 3:	0.00	0.0	0.00
Result quality	Good				

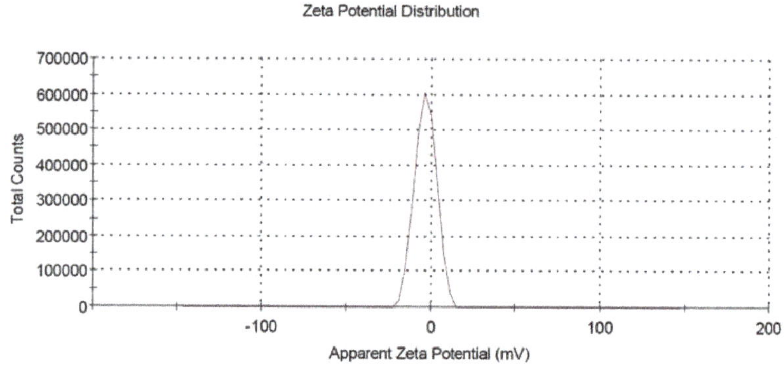

Figure 8. Zeta potential distribution for calcium nanoparticles synthesized with salicylic acid.

4.5. Investigation of CaNP Properties via Transmission Electron Microscopy (TEM)

A transmission electron microscope (JEOL TEM-2100) attached to a CCD camera at a rapid charge of 200 kV was used to determine the size, shape, molecular structure, and morphological data of the acquired CaNPs. Every component of the integrated CaNPs was created by suspending the case on carbon-coated copper networks, allowing allowing dissolvable particles to pass continuously while capturing TEM images (Figure 9). The TEM images were provided by the Central Lab.

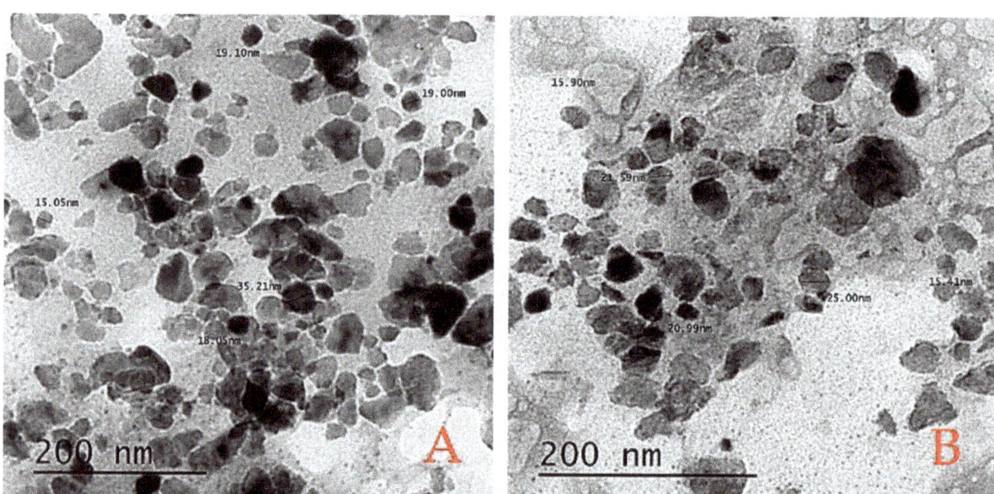

Figure 9. (**A**) The CaNPs had particle sizes ranging from 15.05 to 35.21 nm, while CaNPs containing salicylic acid had particle sizes ranging from 15.90 to 25.00 nm. The particles were mostly spherical, with a few tetragonal particles thrown in for good measure. (**B**) The CaNPs had more aggregated particles than CaNPs containing salicylic acid particles.

4.6. CaNP Application Routine

Cucumber fruits were subjected to four different treatments at the same time. The CaNPs treatments included CaNPs-SA 0 mM, CaNPs-SA 1 mM, and CaNPs-SA 2 mM, as well as a control treatment. Fruits were immersed in the CaNPs-SA treatments for 30 min and then stored for 14 days in ambient conditions (20 ± 1 °C, 60% relative humidity).

4.7. Physical Quality Determination

Weight loss (WL) percentage was determined as the percentage loss of initial weight at harvest time and was calculated by the following formula: $weight\ loss = \frac{WLi-WLs}{WLi} * 100$, where WL_i is the initial weight and WL_s is the weight at a sampling period [48,49].

Cucumber firmness was estimated by utilizing an Instron Universal Testing Machine (Model 4411, Instron, MA, USA) on both sides of the cucumbers in various places on the fruit. A plunger (6 mm diameter) was operated to penetrate the tissues of the skin to the pulp at a depth of 5 mm, and the firmness was measured and presented in N [50].

The cucumber skin color hue was determined using the method described in [51]. The color outline of the fruit was observed with a colorimeter (Minolta CR-300) using the L* a* b* interoperative system proposed by the Commission Internationale de I′ Eclairage (CIE). The assayed color variable (hue angle h°) was verified by applying the equation b*/a*, and the resulting data were represented as recorded in the Minolta camera.

Chlorophyll A and B were determined using N, N-dimethylformamide (DMF) solvent rather than $(CH_3)_2CO$ [52]. The extraction was held at 4 °C for 16 h to allow the DMF to eliminate the pigment from the cucumber skin samples [28]. Finally, the extractions were centrifuged for 10 min at 10,000× g and measured spectrophotometrically at wavelengths of 663.8 nm (ChA) and 646.8 nm (Ch B); The results were reportedin mg 100 g^{-1} FW.

4.8. Cell Wall Enzyme Activities

One gram of cucumber fruit was homogenized with 20 mM Tris-HCl buffer (pH 7). After that, the mixture was centrifuged at 16,000 rpm for 6 min at 4 °C while chilling. Polygalacturonase (PG), cellulase (CEL), and xylanase (XLN) were measured in the filtered sample, which was kept at 20 °C.

In a total volume of 1 mL, we combined polygalacturonic acid and an appropriate amount of enzyme extraction. The addition of the substrate kicked off the process. For one hour, the reaction mixture was maintained at 37 °C. As a result, 500 L of dinitro-salicylic acid reagent was combined and cooked for 10 min in a water bath. The sample was cooled suddenly and the temperature in the lab was attained. With a spectrophotometer, the activities were assessed at 560 nm for PG and XLN and 540 nm for CEL [53].

The lowering of the endpoint of the carboxymethyl cellulose reaction was used to assess cellulase (CEL, EC: 3.2.1.4) enzyme activity [53]. One gram of cucumber was mashed with 20 mM of pH 7.0 Tris-HCL buffer. The mixture was centrifuged for 5 min at 14,000× g while being cooled at 4 °C. To consider CEL at 450 nm, the reaction supernatant was stored at 20 °C. As a measure of the chemical, one unit of compound action was transmitted.

The efficiency of pectinase (PT, EC: 3.2.1.15) was determined [54], and the extraction was performed utilizing the technique outlined in [55]. A mixture of 500 L of polygalacturonic acid (0.36 percent *w/v*) with 0.05 M Tris-HCL buffer (pH 8.5), 200 L of 4 mM $CaCl_2$, 500 L of chemicals, and 500 L water was used to test the enzyme's performance. For 3 h, the response mixture was kept at 36 °C. The enzyme was identified by reading the absorption at 232 nm. This method was also used to add up the solvent protein content in the chemical concentrate. The enzyme's specific activity was measured in milligrams per g^{-1} of protein.

4.9. Lipid Peroxidation and Ion Leakage Percentage

A 2.5 g sample of cucumber pulp was mashed in a mortar with 25 mL of 5% (*w/v*) metaphosphoric acid and 500 L of 2% (*w/v*) butylated hydroxytoluene in ethanol and finally homogenized to determine malondialdehyde (MDA) as a terminative of lipid peroxidation [56]. The standard curve was calibrated using various concentrations of 1,1,3,3-

tetraethyoxypropane (Sigma) in the range of 0–2 mM (TBARS), which was equivalent to 0–1 mM malondialdehyde (MDA). During the acid warming phase of the test, the 1,1,3,3-tetraethyoxypropane is stoichiometrically converted to MDA [57]. MDA concentrations are used to measure the amount of TBARS present.

Five cucumber pulp discs (5 mm in diameter) were soaked three times in normal water to remove any ions before being placed in 10 mL 0.4 M mannitol alcohol in demineralized water at room temperature for three hours. A conductivity meter was used to check the ion conductivity (EC). All samples were boiled in water for 30 min to kill the cucumber tissue cells before being kept cold in the lab. The EC was then measured again, and the samples' relative electrolyte leakage was computed using the following equation: ion leakage percentage = [EC after boiling sample—EC after 3-h -/EC after 3-h] × 100 [56].

4.10. $O_2{}^{\bullet -}$ and H_2O_2 Free Radical Production Rate

One gram of cucumber sample was combined with 3 mL of potassium phosphate buffer (50 mM, pH 7.8) under cooling at 4 °C. The reagent was also combined with poly-vinyl-pyrrolidone (PVP 1% *w/v*), then the mixtures were centrifuged at 11,000 rpm at 4 °C for 15 min. The $O_2{}^{\bullet -}$ generation rate was measured by examining the NO_2 age from hydroxylamine in the presence of $O_2{}^{\bullet -}$ [45]. A standard bend with NO_2 was utilized to verify the $O_2{}^{\bullet -}$generation rate from the response of $O_2{}^{\bullet -}$ to hydroxylamine. The $O_2{}^{\bullet -}$product was presented in nmol min^{-1} g^{-1} FW.

For H_2O_2 determination, a cucumber sample (1 g) was mixed with 5 mL of 100% $(CH_3)_2CO$ then centrifuged at 15,000× *g* for 15 min at 4 °C. One milliliter of the abstraction was mixed with 0.1 mL of 5% Ti $(SO4)_2$ and 0.2 mL of NH_4OH solution. The titanium-peroxide compound was deposited, and the deposit was broken down in 4 mL of 2 M H_2SO_4 after centrifugation at 5000× *g* for 25 min; then, it was quantified on a spectrophotometer at 415 nm. The H_2O_2 matter was calculated from a standard bend balanced likewise and the fixation presented in ηmol g^{-1} FW [58].

4.11. Statistical Analysis

The Software package for co-state Version 6.303 was used to analyze the data (789 light-house Ave PMB 320, Monterey, CA 93940, USA) by taking the mean of the two seasons and using a two-way analysis of variance with the calcium nanoparticle/salicylic acid treatments and storage period (days) as factors. Duncan's multiple range test was used to compare the means of all tested parameters at a $p < 0.05$ level.

5. Conclusions

SA was found to reduce the activity of cell wall degradation enzymes, which not only scavenged H_2O_2 but also reduced $O_2{}^{\bullet -}$ generation in this study. As a result, it is recommended that cucumbers be treated with nano calcium particles mixed with salicylic acid at a concentration of 2 mmol L^{-1} throughout their shelf life.

Author Contributions: Conceptualization, M.F.M.A. and K.M.; Data curation, M.F.M.A., L.A.A., M.A.A., S.I. and S.M.A.D.; Formal analysis, K.M.; Funding acquisition, M.F.M.A. and M.H.M.; Investigation, M.H.M.; Methodology, M.F.M.A., L.A.A., S.I. and S.M.A.D.; Project administration, M.H.M., M.A.A. and S.I.; Resources, L.A.A., M.A.A. and S.M.A.D.; Software, K.M.; Supervision, M.A.A. and S.I.; Validation, L.A.A. and K.M.; Visualization, M.H.M., K.M. and S.M.A.D.; Writing—original draft, M.F.M.A., M.H.M. and K.M.; Writing—review and editing, L.A.A., M.A.A., S.I. and S.M.A.D. All authors will be informed about each step of manuscript processing. All authors have read and agreed to the published version of the manuscript.

Funding: King Saud University (Riyadh, Saudi Arabia) funded this research through the Researchers Supporting Project number (RSP-2021/406).

Institutional Review Board Statement: Not applicable.

Informed Consent Statement: Not applicable.

Data Availability Statement: Data is contained within the article.

Conflicts of Interest: The authors declare no conflict of interest.

Sample Availability: Samples of the compounds are available from the authors.

References

1. FAOSTAT. Source FAOSTAT. 2019. Available online: http://www.fao.org/faostat/en/#data/QC (accessed on 7 May 2022).
2. Lucier, G.; Jerardo, A. Vegetables and Melons Outlook/VGS-319. In *Economic Research Service*; USDA: Washington, DC, USA, 2007.
3. Smith, D.L.; Stommel, J.R.; Fung, R.W.M.; Wang, C.Y.; Whitaker, B.D. Influence of cultivar and harvest method on postharvest storage quality of pepper (*Capsicum annuum* L.) fruit. *Post. Biol. Technol.* **2006**, *42*, 243–247. [CrossRef]
4. Maguire, K.M.; Banks, N.H.; Lang, A. Sources of variation in water vapour permeance of apple fruit. *Post. Biol. Technol.* **1999**, *17*, 11–17. [CrossRef]
5. Cameron, K.D.; Teece, M.A.; Smart, L.B. Increased accumulation of cuticular wax andexpression of lipidtransfer protein in response to periodic drying events in leaves of tree tobacco. *Plant Physiol.* **2006**, *140*, 176–183. [CrossRef] [PubMed]
6. Mamrutha, H.M.; Mogili, T.; Jhansi Lakshmi, K. Leaf cuticular wax amount and crystal morphology regulate postharvest water loss in mulberry (*Morus* species). *Plant Physiol. Biochem.* **2010**, *48*, 690–696. [CrossRef]
7. Rojas-Argudo, C.; Erez-Gago, M.B.P.; Del R'ıo, M.A. Postharvest quality of coated cherries cv. 'Burlat' as affected by coating composition and solids content. *Food Sci. Technol. Int.* **2005**, *11*, 417–424. [CrossRef]
8. Hodges, D.M.; Lester, G.E.; Munro, K.D.; Toivonen, P.M.A. Oxidative stress: Importance for postharvest quality. *HortScience* **2004**, *39*, 924–929. [CrossRef]
9. Porat, R.; Weiss, B.; Cohen, L.; Daus, A.; Aharoni, N. Reduction of postharvest rind disorders in citrus fruit by modified atmosphere packaging. *Postharvest Biol. Technol.* **2004**, *33*, 35–43. [CrossRef]
10. Singh, C.; Sharma, H.K.; Sarkar, B.C. Influence of process conditions on the mass transfer during osmotic dehydration of coated pineapple samples. *J. Food Process. Preserv.* **2010**, *34*, 700–714. [CrossRef]
11. Saini, C.S.; Sharma, H.K. Effect of pectin coating on colour and quality of dehydrated pineapple during storage. *Asian J. Dairy Food Res.* **2016**, *35*, 120–129. [CrossRef]
12. Jiang, Y.; Li, Y.; Jiang, E. Effects of chitosan on shelf life of cold-stored litchi fruit at ambient temperature. *LWT Food. Sci. Technol. Int.* **2005**, *38*, 757–761. [CrossRef]
13. Babalar, M.; Asghari, M.; Talaei, A.R.; Khosroshahi, A. Effect of pre- and postharvest salicylic acid treatment on ethylene production, fungal decay and overall quality of selva strawberry fruit. *Food Chem.* **2007**, *105*, 449–453. [CrossRef]
14. Kmita, A.; Hutera, B.; Olejnik, E.; Janas, A. Effect of water glass modification with nanoparticles of zinc oxide on selected physical and chemical properties of binder and mechanical properties of sand mixture. *Arch. Foundry Eng.* **2012**, *12*, 37–40. [CrossRef]
15. Ali, I. Nano anti-cancer drugs: Pros and cons and future perspectives. *Curr. Cancer Drug Targets* **2011**, *11*, 131–134. [CrossRef] [PubMed]
16. Lo'ay, A.A.; Ameer, N.M. Performance of calcium nanoparticles blending with ascorbic acid and alleviation internal browning of 'Hindi Be-Sennara' mango fruit at a low temperature. *Sci. Hortic.* **2019**, *254*, 199–207. [CrossRef]
17. Zavala, I.F.A.; Wangs, Y.; Wang, C.Y.; Aguilar, G.A.A. Effects of storage temperatures on antioxidant capacity and aroma compounds in strawberry fruit. *LWT-Food Sci. Technol.* **2004**, *37*, 687–695. [CrossRef]
18. Agamy, R.A.; Elsayed, E.H.; Tarek, H.T. Acquired resistant motivated by salicylic acid applications on salt stressed tomato (*Lycopersicon esculentum* Mill.). *Am. -Eurasian J. Agric. Environ. Sci.* **2013**, *13*, 50–57. [CrossRef]
19. Dat, I.; Vandenabeele, S.; Vra Nova, E.; Vanmontagu, M.; Inze, D.; Vanbreusegem, F. Dual action of the active oxygen species during plant stress responses. *Cell Mol. Life Sci.* **2000**, *57*, 779–795. [CrossRef]
20. Orabi, S.; Dawood, M.; Salman, S. Comparative study between the physiological role of hydrogen peroxide and salicylic acid in alleviating the harmful effect of low temperature. *Sci. Agric.* **2015**, *9*, 49–59. [CrossRef]
21. Khan, W.; Prithiviraj, B.; Smith, D. Photosynthetic responses of corn and soybean to foliar application of salicylates. *J. Plant Physiol* **2003**, *160*, 485–492. [CrossRef]
22. Hayat, Q.; Hayat, S.; Irfan, M.; Ahmad, A. Effect of exogenous salicylic acid under changing environment: A review. *Environ. Exper. Bot.* **2010**, *68*, 14–25. [CrossRef]
23. Srivastava, M.K.; Dwivedi, U.N. Delayed ripening of banana fruit by salicylic acid. *Plant Sci.* **2000**, *158*, 87–96. [CrossRef]
24. Zhang, Y.; Chen, K.; Zhang, S.; Ferguson, I. The role of salicylic acid in postharvest ripening of kiwifruit. *Postharvest Biol. Technol.* **2003**, *28*, 67–74. [CrossRef]
25. Fattahi, J.; Fifaii, R.; Babri, M. Postharvest quality of kiwifruit (*actinidia deliciosa* cv. hayward) affected by pre-storage application of salicylic acid. *S. W. J. Hortic. Biol. Environ.* **2010**, *1*, 175–186. Available online: http://anucraiova.3x.ro/swjhbe/index.html (accessed on 7 May 2022).
26. Yao, H.; Tian, S.H. Effects of pre- and post-harvest application of salicylic acid or methyl jasmonate on inducing disease resistance of sweet cherry fruit in storage. *Postharvest Biol. Technol.* **2005**, *35*, 253–262. [CrossRef]
27. Ding, C.K.; Wang, C.Y.; Gross, K.C.; Smith, D.I. Jasmonate and salicylate induce the expression of pathogenesis-related-protein genes and increase resistance to chilling injury in tomato fruit. *Planta* **2002**, *214*, 895–901. [CrossRef] [PubMed]

28. Lo'ay, A.A.; Taher, M.A. Influence of edible coatings chitosan/PVP blending with salicylic acid on biochemical fruit skin browning incidence and shelf life of guava fruits cv. 'Banati'. *Sci. Hortic.* **2018**, *235*, 424–436. [CrossRef]
29. Pasanphan, W.; Buettner, G.R.; Chirachanchai, S. Chitosan gallate as a novel potential polysaccharide antioxidant: An EPR study *Carbohydr. Res.* **2010**, *345*, 132–140. [CrossRef]
30. Tareen, M.J.; Abbasi, N.A.; Hafiz, I.A. Postharvest application of salicylic acid enhanced antioxidant enzymes activity and maintanied quality of peach cv. 'Flordaking' fruit during storage. *Sci. Hortic.* **2012**, *142*, 221–228. [CrossRef]
31. El-Ezz, S.F.A.; Al-Harbi, N.A.; Al-Qahtani, S.M.; Allam, H.M.; Abdein, M.A.; Abdelgawad, Z.A. A Comparison of the Effects of Several Foliar Forms of Magnesium Fertilization on 'Superior Seedless' (*Vitis vinifera* L.) in Saline Soils. *Coatings* **2022**, *12*, 201. [CrossRef]
32. Sun, X.; Wang, Z.; Kadouh, H.; Zhou, K. The antimicrobal, mechanical, physical and structureal properies of chitosan-gallic acid film. *LWT Food Sci. Techol.* **2014**, *57*, 83–89. [CrossRef]
33. Foyer, C.H.; Ruban, A.V.; Noctor, G. Viewing oxidative stress through the lens of oxidative signaling rather than damage. *Biochem. J.* **2017**, *474*, 877–883. [CrossRef] [PubMed]
34. Sayyari, M.; Babalar, M.; Kalantari, S.; Martínez-Romero, D.; Guillén, F.; Serrano, M.; Valero, D. Vapour treatments with methyl salicylate or methyl jasmonate alleviated chilling injury and enhanced antioxidant potential during postharvest storage of pomegranates. *Food Chem.* **2011**, *124*, 964–970. [CrossRef]
35. Lo'ay, A.A.; Elgammal, R.E.; Alhaithloul, H.A.S.; Alghanem, S.M.; Fikry, M.; Abdein, M.A.; Hikal, D.M. Enhance Fruit Ripening Uniformity and Accelerate the Rutab Stage by Using ATP in 'Zaghloul' Dates during the Shelf Life. *Foods* **2021**, *10*, 2641. [CrossRef] [PubMed]
36. Shoala, T.; Al-Karmalawy, A.A.; Germoush, M.O.; Alshamrani, S.M.; Abdein, M.A.; Awad, N.S. Nanobiotechnological Approaches to Enhance Potato Resistance against Potato Leafroll Virus (PLRV) Using Glycyrrhizic Acid Ammonium Salt and Salicylic Acid Nanoparticles. *Horticulturae* **2021**, *7*, 402. [CrossRef]
37. Youwei, Y.; Yinzhe, R. Grapes preservation using chitosan combined with β−cyclodextrin. *Int. J. Agron.* **2013**, *4*, 1–8. [CrossRef]
38. Lo'ay, A.A.; Mostafa, N.A.; Al-Qahtani, S.M.; Al-Harbi, N.A.; Hassan, S.; Abdein, M.A. Influence of the Position of Mango Fruit on the Tree (Mangifera indica L. CV. 'Zibda') on Chilling Sensitivity and Antioxidant Enzyme Activity. *Horticulturae* **2021**, *7*, 515. [CrossRef]
39. Lo'ay, A.A. Chilling Injury in Mangoes. Ph.D. Thesis, Wageningen University, Wageningen, The Netherlands, 2005; pp. 1–224, ISBN: 90-8504-309-3.
40. Purvis, A.C. Regulatin of oxidative stress in horticultural crops. *HortiScience* **2004**, *39*, 930–932. [CrossRef]
41. Taher, M.A.; Lo'ay, A.A.; Gouda, M.; Limam, S.A.; Abdelkader, M.F.M.; Osman, S.O.; Fikry, M.; Ali, E.F.; Mohamed, S.Y.; Khalil, H.A.; et al. Impacts of Gum Arabic and Polyvinylpyrrolidone (PVP) with Salicylic Acid on Peach Fruit (*Prunus persica*) Shelf Life. *Molecules* **2022**, *27*, 2595. [CrossRef]
42. Champa, W.A.H.; Gill, M.I.S.; Mahajan, B.V.C.; Arora, N.K. Preharvest salicylic acid treatments to improve quality and postharvest life of table grapes (*Vitis vinifera* L.) cv. Flame Seedless. *J. Food Sci. Technol.* **2015**, *52*, 3607–3616. [CrossRef]
43. Roustan, J.P.; Latché, A.; Fallot, J. Inhibition of ethylene production and stimulation of carrot somatic embryogenesis by salicylic acid. *Biol. Plant* **1990**, *32*, 273–276. [CrossRef]
44. Tang, T.; Huang, D.-W.; Zhou, C.-Q.; Li, X.; Xie, Q.-J.; Liu, F.-S. Molecular cloning and expression patterns of copper/zinc superoxide dismutase and manganese superoxide dismutase in Musca domestica. *Gene* **2012**, *505*, 211–220. [CrossRef]
45. Yang, H.; Wu, F.; Cheng, J. Reduced chilling injury in cucumber by nitricoxide and the antioxidant response. *Food Chem.* **2011**, *127*, 1237–1242. [CrossRef] [PubMed]
46. Yugandhar, P.; Savithramma, N. Green synthesis of calcium carbonate nanoparticles and their effects on seed germination and seedling growth of *Vigna mungo* (L.). Hepper cv. *Int. J. Adv. Res.* **2013**, *1*, 89–103. Available online: http://www.journalijar.com/uploads/2013-11-30_140955_59.pdf (accessed on 7 May 2022).
47. Soheyla, H.; Foruhe, Z. Effect of Zeta Potential on the Properties of Nano-Drug Delivery Systems—A Review (Part 2). *Trop. J. Pharm. Res.* **2013**, *12*, 265–273. [CrossRef]
48. Naveena, B.; Genitha, I. Effect of calcium chloride, sodium chloride and lime juice on physico-chemical properties of cucumber. *Int. J. Agric. Sci. Res.* **2017**, *7*, 765–770. [CrossRef]
49. Lo'ay, A.A.; Ismail, H.; Kassem, H.S. The Quality of Superior Seedless Bunches during Shelf Life as Determined by Growth on Different Rootstocks. *Agriculture* **2021**, *11*, 990. [CrossRef]
50. Omoba, O.S.; Onyekwere, U. Postharvest physicochemical properties of cucumber fruits (*Cucumber sativus* L) treated with chitosan-lemon grass extracts under different storage durations. *Afr. J. Biotechnol.* **2016**, *15*, 2758–2766. [CrossRef]
51. Nardello, I.C.; Cantillano, R.F.F.; Seifer, M.; Mello-Farias, P.C.; Malgarim, M.B.; De Oliveira, R.P. Postharvest quality during refrigerated storage of 'Nadorcott' mandarin. *J. Exper. Agric. Inter.* **1994**, *19*, 1–10. [CrossRef]
52. Porra, R.J.; Thompson, W.A.; Kriedemann, P.E. Determination of accurate extinction coefficients and simultaneous equations for assaying chlorophylls a and b extracted with four different solvents: Verification of the concentration of chlorophyll standards by atomic absorption spectroscopy. *Biochim. Biophys. Acta* **1989**, *975*, 384–394. [CrossRef]
53. Miller, G.L. Use of dinitrosalicylic acid reagent for the determination of reducing suga. *Anal. Chem.* **1959**, *31*, 426–429. [CrossRef]
54. Collmer, A.; Reid, J.L.; Mount, M.S. Assay methods for pectic enzymes. In *Methods in Enzymology*; Wood, W.A., Kellogg, S.T., Eds.; Academic Press: San Diego, CA, USA, 1988; Volume 161, pp. 329–335.

55. Payasi, A.; Sanwal, G.G. Pectate lyase activity during ripening of banana fruit. *Phytochemistry* **2003**, *63*, 243–248. [CrossRef]
56. Lo'ay, A.A.; Dawood, H.D. Active chitosan/PVA with ascorbic acid and berry quality of 'Superior seedless' grapes. *Sci. Hortic.* **2017**, *224*, 286–292. [CrossRef]
57. Iturbe-Ormaetxe, I.; Escuredo, P.R.; Arrese-Igor, C.; Becana, M. Oxidative damage in pea plants exposed to water deficit or paraquat. *Plant Physiol.* **1998**, *116*, 173–181. [CrossRef]
58. Xu, M.; Dong, J.; Zhang, M.; Xu, X.; Sun, L. Cold-induced endogenous nitric oxide generation plays a role in chilling tolerance of loquat fruit during postharvest storage. *Postharvest Biol. Technol.* **2012**, *65*, 5–12. [CrossRef]

molecules MDPI

Article

Impacts of Gum Arabic and Polyvinylpyrrolidone (PVP) with Salicylic Acid on Peach Fruit (*Prunus persica*) Shelf Life

Mohamed A. Taher [1], A. A. Lo'ay [2,*] Mostafa Gouda [3,4,*], Safaa A. Limam [5], Mohamed F. M. Abdelkader [6], Samah O. Osman [7], Mohammad Fikry [8], Esmat F. Ali [9], Sayed. Y. Mohamed [7], Hoda A. Khalil [10] Diaa O. El-Ansary [11], Sherif F. El-Gioushy [12], Hesham S. Ghazzawy [13,14,*], Aly M. Ibrahim [7], Mahmoud F. Maklad [15], Mohamed A. Abdein [16,*] and Dalia M. Hikal [17]

1 Agricultural Chemistry Department, Faculty of Agriculture, Mansoura University, El-Mansoura 35336, Egypt; mohamedtaher@mans.edu.eg
2 Pomology Department, Faculty of Agriculture, Mansoura University, El-Mansoura 35336, Egypt
3 College of Biosystems Engineering and Food Science, Zhejiang University, Hangzhou 310058, China
4 Department of Nutrition and Food Science, National Research Centre, Dokki, Giza 12422, Egypt
5 Food Science and Technology Department, Faculty of Agriculture, Assiut University, Assiut 71526, Egypt; limamsafaa@gmail.com
6 Department of Plant Production, College of Food and Agriculture, King Saud University, Riyadh 12372, Saudi Arabia; mohabdelkader@ksu.edu.sa
7 Horticulture Research Institute, Agricultural Research Center, Giza 12619, Egypt; ayatosman012@gmail.com (S.O.O.); sayed_h_11@yahoo.com (S.Y.M.); dr.alyibrahim70@gmail.com (A.M.I.)
8 Department of Agricultural and Biosystems Engineering, Faculty of Agriculture, Benha University, Moshtohor, Toukh 13736, Egypt; moh.eltahlawy@fagr.bu.edu.eg
9 Department of Biology, College of Science, Taif University, Taif 21944, Saudi Arabia; a.esmat@tu.edu.sa
10 Department of Pomology, Faculty of Agriculture (EL-Shatby), Alexandria University, Alexandria 21545, Egypt; hoda.khalil@alexu.edu.eg
11 Precision Agriculture Laboratory, Pomology Department, Faculty of Agriculture (El-Shatby), Alexandria University, Alexandria 21545, Egypt; diaa.elansary@alexu.edu.eg
12 Horticulture Department, Faculty of Agriculture (Moshtohor), Benha University, Moshtohor, Toukh 13736, Egypt; S.F.El-Gioushy@yahoo.com or sherif.elgioushy@fagr.bu.edu.eg
13 Date Palm Research Center of Excellence, King Faisal University, Hofuf 31982, Saudi Arabia
14 Central Laboratory for Date Palm Research and Development, Agriculture Research Center, Giza 12511, Egypt
15 Department of Horticulture, Faculty of Agriculture, Ain Shams University, Cairo 11566, Egypt; mahmoud_maklad@agr.asu.edu.eg
16 Biology Department, Faculty of Arts and Science, Northern Border University, Rafha 91911, Saudi Arabia
17 Nutrition and Food Science, Home Economics Department, Faculty of Specific Education, Mansura University, Mansoura 35516, Egypt; dr.daliahikal@mans.edu.eg
* Correspondence: Loay_arafat@mans.edu.eg (A.A.L.); mostafa-gouda@zju.edu.cn (M.G.); hghazzawy@kfu.edu.sa (H.S.G.); abdeingene@yahoo.com (M.A.A.)

Abstract: Peaches are grown in many Egyptian orchards for local and global fresh market sales. The interior fruit tissue breakdown (IFTB), often resulting in decayed peaches, is a severe problem during marketing. Therefore, to minimize FTB of peaches, in this study, gum arabic (GA) and polyvinylpyrrolidone (PVP) were mixed with different concentrations of salicylic acid (SA) (0, 1, and 2 mM) and were applied as edible coating to extend the shelf life of peach fruits. Mature peaches were selected and harvested when peaches reached total soluble solid content (SSC: 8.5%) and fruit firmness of about 47 N. Fruits were coated and stored at room temperature (26 ± 1 °C and air humidity 51 ± 1%) for 10 days during two seasons: 2020 and 2021. Fruit coated with GA/PVP-SA 2 mM showed a significant ($p < 0.05$) inhibition in degrading enzyme activities (CWDEs), such as lipoxygenase (LOX), cellulase (CEL), and pectinase (PT), compared to uncoated and coated fruits during the shelf-life period. Hence, cell wall compartments were maintained. Consequently, there was a reduction in browning symptoms in fruits by inhibiting polyphenol oxidase (PPO) and phenylalanine ammonia-lyase (PAL) activities. Thus, the fruit skin browning index showed almost no symptoms. The lipid peroxidation process and ionic permeability declined as well. The result suggests that, by applying GA/PVP-SA 2 mM as an edible coating, fruit tissue breakdown can be minimized, and the shelf life of peach can be extended up to 10 days without symptoms of tissue breakdown.

Keywords: peach; shelf-life; cell-wall-degrading enzymes; plant tissue; gum arabic

1. Introduction

The peach fruit (*Prunus persica* L. cv, 'Early Sweet'), which belongs to the family Rosaceae, is one of the most popular fruits in the world and, in particular, in the Egyptian market due to its nutritive value and characteristic flavor [1]. Moreover, this fruit contains considerable amounts of bioactive pigments, such as anthocyanins, carotenoids, lutein, and β-cryptoxanthin [2]. Peach fruits lose water and wilt quickly as they progress through the ripening cycle. Therefore, the short postharvest life of peach fruit makes it unmarketable due to the extreme breakdown of interior tissues accompanied by microbial infections [3]. Interior tissue breakdown in peach fruit, indicated by flesh browning, is characterized by elevated neutral sugar and low amounts of cellulose and pectin, as well as diminished activity of pectin-hydrolyzing enzymes and cation binding, mostly calcium, in the cell wall [4]. Cold storage is proposed as the most appropriate technique to slow down the decay processes and preserve the fruit quality. However, peach fruits are sensitive to low storage temperatures, which limit their storage period [5].

Currently, some edible polymers are being extensively applied alone to prolong the shelf life and quality attributes of edible fruits [4,5]. In the case of peach fruit, the desirable effects of different *Aloe* gels on delaying ethylene production, ripening index, color development, and weight loss during storage of peach fruits at ambient temperature have been reported [6]. Additionally, *Aloe vera* gel alone has been documented as a suitable edible film for peach fruits regarding shelf life [7].

The blending of biodegradable polymers is one of the most valuable techniques to produce a new material with mechanically desirable properties in comparison with the individual polymers. In this respect, the prolonged shelf life of guava fruit by a cashew gum/carboxymethyl cellulose biopolymer blend has been reported [8]. The incorporation of blends of chitosan/PVA biopolymers and oxalic or ascorbic acids alone were, respectively, reported to alleviate skin browning in bananas [9] and to minimize the activities of cell-wall-degrading enzymes (CWDEs) of 'superior seedling' grapes during storage [10,11]. Furthermore, these blends inhibit the fungal infections [12] and preserve the quality of the fruit [13].

Gum arabic (GA) is one of the most common polysaccharides and is naturally extracted from the bark of *Acacia senegal*. The Food and Agriculture Organization of the United Nations (FAO) has permitted GA as a safe additive coating in food industries [14]. Moreover, a blended coating containing GA and chitosan was also stated to enhance fruit quality in banana [15,16]. Structurally, arabinogalactan is the main component of GA (80–90%) [17]. The characteristic properties of GA have explained its diverse applications, i.e., emulsification [18], stabilization [19], and micro-encapsulation [20]. GA is widely employed in the postharvest processing of edible fruits [21–23]. Polyvinylpyrrolidone (PVP) is a safe nonionic amorphous polymer with a high level of solubility in water [24]. PVP, with its useful properties, such as water solubility, absence of toxicity, film development, and adhesive power, is one of the best hopeful polymers for nanogels research [25]. PVP has been considered to be a prospective polymer with a great film-forming capacity for probable application in the production of coatings, cosmetics, detergents, plastics, medicine, and pharmaceuticals [26,27]. The United States Food and Drug Administration (FDA) has permitted PVP for various applications [28]. Acceptable daily intake of PVP has been documented as 0–50 mg/kg/day [29]. Food applications of PVP, including coating for fresh citrus fruits, clarification of beverages, binder for vitamin and mineral concentrates, binder for synthetic sweeteners, and dehydration of aqueous foods, such as orange and tomato juices, have been documented [30].

Salicylic acid (SA; also known as 2-hydroxy benzoic acid) is an endogenous growth regulator and a signal molecule that is critical for the induction of resistance to biotic and abiotic stress. In plants, it exists as a free phenolic acid and as conjugated forms that are constructed by hydroxylation, glycosylation, or methylation of the aromatic ring [31]. Earlier studies have stated the profits of preharvest and postharvest treatments with SA on different fruit quality features, such as ascorbic acid content, greater weight, and firmness in peach fruit [31] and grape fruit [32]; lower level of degradation of carotenoids in grape tissue color [33] and better chroma index; and higher total soluble solids, bioactive ingredients, and antioxidant activities and enhanced activity of some antioxidant enzymes in sweet cherry fruit [34,35]. Much research has stated the higher contents of total polyphenols and flavonoids in some SA-treated fruits, such as sweet cherries [34], peaches [36], and apples [37]. Moreover, numerous studies have stated the benefits of postharvest applications with SA in polymer coatings, such as chitosan/PVP in guava [31] and chitosan enriched by nanosized titanium dioxide particles in blackcurrant fruit [33].

Regarding public doubts about the undesirable effects of synthetic fungicides on human health and the environment, there is continuing research into new substitutes for the application of synthetic chemicals. One of the alternatives might be the use of SA, which has revealed antifungal properties on some fruits and other plants [38]. For instance, Babalar et al. [38] stated the effectiveness of SA against the decay caused by *Botrytis cinerea* in strawberry fruit. It is also recognized that SA has fungicidal properties on the brown rot of sweet cherry fruit, which is caused by *Monilinia fructicola* [39].

There are no available data in the literature concerning the use of a biopolymer blend of GA/PVP supplemented with salicylic (SA) acid as an edible coating. Thus, the present study aimed to evaluate the ability of GA/PVP/SA edible coating formulations in minimizing the incidences of tissue breakdown in peach fruits' 'Early Sweet' during the trading period.

2. Results

2.1. Fourier Transform Infrared Analysis (FT-IR)

Figure 1 shows FT-IR spectra of pure SA, GA, PVP, and the blends of GA/PVP and GA/PVP–SA 2 mM. The spectra showed characteristic bands of vibrations of the functional groups formed in the prepared blends. The IR spectrum of salicylic acid (Figure 1) reveal the absorption bands at 3238, 3063, 1659, 1612, 1577, 1483, 1443, and 1297 cm^{-1}. Meanwhile, the peaks at 3422, 2927, 1654, 1429, 1030, and 776 cm^{-1} were noted for GA. The IR spectrum of PVP shows the characteristic peaks observed at 3450, 1656, 1435, 1289, and 845 cm^{-1}. Moreover, the characteristic bands at 3451, 2926, 1294, and 776 cm^{-1} were found in the GA/PVP blend (Figure 1). The FT-IR spectrum of GA/PVP–SA 2 mM has prominent bands at 3528, 3240, 1660, 1612, 1483, 1296, and 760 cm^{-1} wavenumbers.

2.2. Evaluation of Fruit Properties: Water Loss %, Skin Browning Index (SBI), Hue Angle, and Firmness (N)

GA/PVP blends as edible coatings possessed significant effects on weight loss of peach fruits compared with control uncoated fruits (Table 1). The GA/PVP–SA 2 mM coating treatment recorded the lowest percentage of water loss (%) throughout the storage period. The skin browning index of control uncoated fruits recorded a slow raise until the sixth day, followed by a sharp increase and, therefore, reached the highest value (3.75) at the end of the storage period (Table 1). Meanwhile, its values did not alter throughout the experiment period in fruits treated with GA/PVP/SA composite blends. An acceptable low level of SBI was also obtained when peach fruit was treated with a GA/PVP blend.

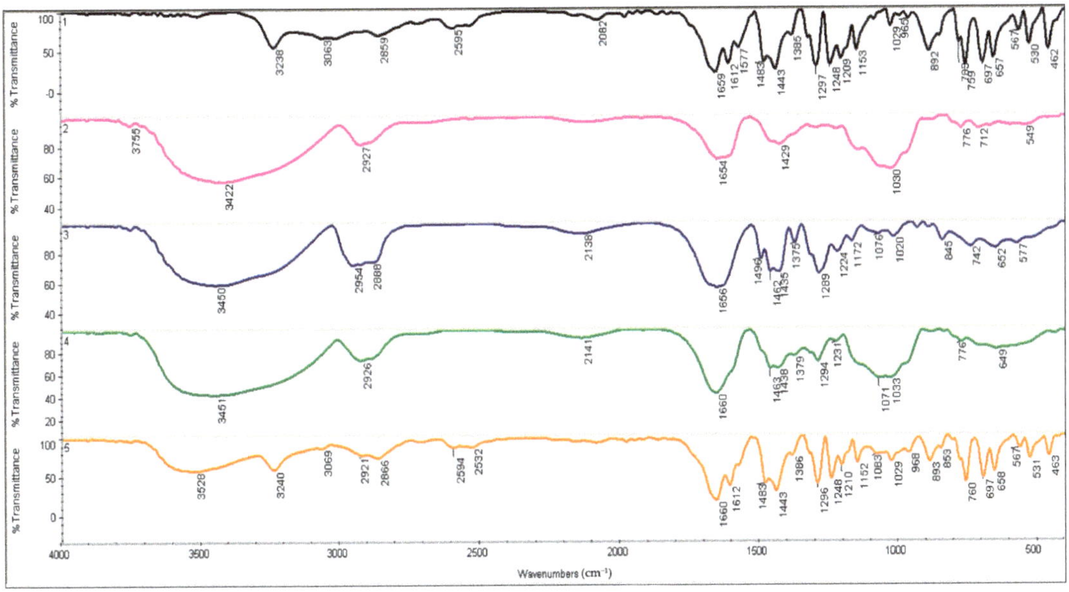

Figure 1. FTIR spectra of (1) SA, (2) GA, (3) PVP, (4) GA/PVP, and (5) GA/PVP/SA 2 mM. FT-IR, Fourier transform infrared spectroscopy; SA, salicylic acid; GA, gum arabic; PVP, polyvinylpyrrolidone.

Table 1. Effect of different treatments of gum arabic (GA) and polyvinylpyrrolidone (PVP) blending with salicylic acid (SA) at different concentrations as a mixture biopolymer coating treatment (GA/PVP–SA) on physical properties of peach fruit 'Early Sweet' during shelf life at ambient air (29 ± 1 °C and air humidity 51%) during two seasons (2020 and 2021).

Treatments	Shelf-Life Period (Days)					
	D_0	D_2	D_4	D_6	D_8	D_{10}
Water loss %						
Control	0.00 ± 0.000	3.14 ± 0.017 ^{a}	11.97 ± 1.097 ^{a}	18.47 ± 0.784 ^{a}	27.35 ± 0.784 ^{a}	31.30 ± 0.708 ^{a}
GA/PVP–SA 0 mM	0.00 ± 0.000	3.06 ± 0.012 ^{b}	9.46 ± 0.309 ^{b}	12.46 ± 0.652 ^{b}	19.85 ± 0.311 ^{b}	22.41 ± 0.584 ^{b}
GA/PVP–SA 1 mM	0.00 ± 0.000	3.02 ± 0.014 ^{c}	7.86 ± 0.236 ^{c}	9.67 ± 0.287 ^{c}	16.71 ± 0.898 ^{c}	18.28 ± 0.545 ^{c}
GA/PVP–SA 2 mM	0.00 ± 0.000	2.98 ± 0.008 ^{d}	6.66 ± 0.193 ^{c}	8.68 ± 0.193 ^{d}	12.42 ± 0.506 ^{d}	14.44 ± 0.578 ^{d}
Skin browning index						
Control	1.00 ± 0.000	1.02 ± 0.006 ^{a}	1.16 ± 0.005 ^{a}	1.60 ± 0.031 ^{a}	1.99 ± 0.003 ^{a}	3.75 ± 0.122 ^{a}
GA/PVP–SA 0 mM	1.00 ± 0.000	1.00 ± 0.000 ^{b}	1.02 ± 0.014 ^{b}	1.10 ± 0.008 ^{b}	1.19 ± 0.008 ^{b}	1.93 ± 0.061 ^{b}
GA/PVP–SA 1 mM	1.00 ± 0.000	1.00 ± 0.000 ^{b}	1.00 ± 0.000 ^{c}	1.00 ± 0.000 ^{c}	1.01 ± 0.005 ^{c}	1.08 ± 0.015 ^{c}
GA/PVP–SA 2 mM	1.00 ± 0.000	1.00 ± 0.000 ^{b}	1.00 ± 0.000 ^{c}	1.00 ± 0.000 ^{c}	1.00 ± 0.000 ^{c}	1.02 ± 0.005 ^{c}
Fruit color (hue angle)						
Control	5.81 ± 0.316 ^{a}	8.67 ± 0.299 ^{d}	12.28 ± 0.191 ^{a}	12.80 ± 0.507 ^{d}	9.59 ± 0.485 ^{d}	9.29 ± 0.024 ^{d}
GA/PVP–SA 0 mM	5.81 ± 0.316 ^{a}	9.85 ± 0.141 ^{c}	13.09 ± 0.054 ^{c}	14.19 ± 0.193 ^{c}	13.46 ± 0.277 ^{c}	12.57 ± 0.2137 ^{c}
GA/PVP–SA 1 mM	5.81 ± 0.316 ^{a}	10.38 ± 0.175 ^{b}	13.91 ± 0.200 ^{b}	15.21 ± 0.153 ^{b}	16.07 ± 0.256 ^{b}	15.26 ± 0.021 ^{b}
GA/PVP–SA 2 mM	5.81 ± 0.316 ^{a}	11.43 ± 0.158 ^{a}	15.61 ± 0.300 ^{a}	16.41 ± 0.515 ^{a}	16.88 ± 0.106 ^{a}	17.96 ± 0.229 ^{a}
Fruit firmness (*N*)						
Control	47.68 ± 0.655 ^{a}	39.35 ± 0.498 ^{ef}	32.38 ± 1.387 ^{i}	29.50 ± 0.329 ^{j}	25.71 ± 0.472 ^{k}	19.30 ± 0.572 ^{l}
GA/PVP–SA 0 mM	47.68 ± 0.655 ^{a}	40.63 ± 0.304 ^{de}	38.03 ± 0.342 ^{f}	35.18 ± 0.536 ^{h}	32.19 ± 0.562 ^{i}	29.59 ± 0.367 ^{j}
GA/PVP–SA 1 mM	47.68 ± 0.655 ^{a}	44.25 ± 0.351 ^{b}	42.81 ± 0.682 ^{c}	39.43 ± 0.625 ^{e}	36.55 ± 0.664 ^{g}	31.62 ± 0.888 ^{i}
GA/PVP–SA 2 mM	47.68 ± 0.655 ^{a}	46.70 ± 0.894 ^{a}	44.41 ± 0.385 ^{b}	42.59 ± 0.628 ^{c}	40.80 ± 0.572 ^{d}	39.47 ± 0.398 ^{de}

Means in the same column that have different letter(s) are significantly different using Duncan's multiple range test at $p < 0.05$ and ±SE (n = 3) replicates. Skin browning spot index was used to determine skin browning (1 = healthy; 2 = slightly brown; 3 = moderately brown; 4 = severe brown; and 5 = fully brown).

GA/PVP blends had a significant impact on the hue angle (color appearance parameter) of peach fruits compared with control uncoated fruits (Table 1). Throughout the experiment, fruits treated with GA/PVP blends developed higher hue angle values in comparison with the non-coated fruits. The color appearance parameter (ho) of fruits ultimately drops; however, the level of decrease in the fruits coated with GA/PVP blends is much lower than the uncoated fruits. Moreover, GA/PVP–SA 2 mM treatment displayed the highest ho value (17.96) on the 10th day of the storage period (Table 1).

Results indicated that GA/PVP coating blends have a significant impact on peach fruit firmness compared with control uncoated fruits (Table 1). The fruit firmness gradually decreased all across the storage period; however, the grade of decrease in the fruits coated with GA/PVP blends is much lower than untreated fruits. The GA/PVP blend enriched with 2 mM salicylic acid in this study resulted in the lowest loss of firmness of peach fruits, which was noticeably more effectual throughout the short period (10 days) of storage at 25 °C (Table 1).

2.3. The Fruit Skin Browning Variables: Total Phenol (TPs), Flavonoids (FLs), and Browning Enzyme Activities

During 10 days of storage at 25 °C, the contents of TPs and TFs of uncoated fruits were significantly lower than those of GA/PVP treatments (Table 2). Noticeably, the phenolic contents regularly diminished in coated and uncoated fruits throughout the shelf-life period. Moreover, the GA/PVP–SA 2 mM treatment recorded the highest contents of TPs and TFs during the 10 days of storage time of peach fruits. The treatment of GA/PVP–SA 2 mM showed the highest TP and FL contents at the end of the experiment (TP, 79.89; FL, 23.41 mg 100 g^{-1}), while the control fruits displayed a progressive decline in phenolic contents on the 10th day (TP, 55.90; FL, 12.38 mg 100 g^{-1} FW), respectively.

Table 2. Effect of different treatments of gum arabic (GA) and polyvinylpyrrolidone (PVP) blending with salicylic acid (SA) at different concentrations as a mixture biopolymer coating treatment (GA/PVP–SA) on skin browning elements of peach fruit 'Early Sweet' during shelf life at ambient air (29 ± 1 °C and air humidity 51%) during two seasons (2020 and 2021).

Treatments	Shelf-Life Period (Days)					
	D_0	D_2	D_4	D_6	D_8	D_{10}
Total phenol content (TP; mg 100 g^{-1} FW)						
Control	86.58 ± 0.858 ^{ab}	79.84 ± 0.896 ^{de}	77.67 ± 0.360 ^{fg}	66.80 ± 0.378 ^{j}	60.86 ± 0.473 ^{k}	55.90 ± 1.815 ^{l}
GA/PVP–SA 0 mM	86.58 ± 0.858 ^{ab}	83.55 ± 0.619 ^{c}	79.07 ± 0.330 ^{ef}	75.39 ± 0.641 ^{h}	70.99 ± 0.291 ^{i}	67.56 ± 0.636 ^{j}
GA/PVP–SA 1 mM	86.58 ± 0.858 ^{ab}	85.57 ± 0.569 ^{b}	81.57 ± 0.715 ^{d}	79.56 ± 0.402 ^{ef}	77.63 ± 0.498 ^{fg}	76.74 ± 0.531 ^{gh}
GA/PVP–SA 2 mM	86.58 ± 0.858 ^{ab}	87.68 ± 0.200 ^{a}	86.43 ± 0.646 ^{ab}	81.66 ± 0.536 ^{cd}	80.12 ± 0.360 ^{de}	79.89 ± 0.564 ^{de}
Flavonoids content (FL; mg 100 g^{-1} FW)						
Control	33.69 ± 1.035 ^{a}	29.47 ± 0.691 ^{d}	26.44 ± 0.543 ^{e}	17.34 ± 0.906 ^{j}	15.40 ± 0.601 ^{k}	12.38 ± 0.590 ^{l}
GA/PVP–SA 0 mM	33.69 ± 1.035 ^{a}	30.91 ± 0.131 ^{b}	29.38 ± 0.306 ^{d}	21.41 ± 0.694 ^{h}	19.61 ± 0.528 ^{i}	16.92 ± 0.344 ^{j}
GA/PVP–SA 1 mM	33.69 ± 1.035 ^{a}	31.53 ± 0.117 ^{b}	30.57 ± 0.129 ^{bc}	26.44 ± 0.543 ^{e}	24.85 ± 0.626 ^{f}	21.83 ± 0.294 ^{h}
GA/PVP–SA 2 mM	33.69 ± 1.035 ^{a}	33.03 ± 0.389 ^{a}	31.36 ± 0.317 ^{b}	29.59 ± 0.442 ^{cd}	26.40 ± 0.528 ^{e}	23.41 ± 0.463 ^{g}
Phenylalanine ammonia-lyase activity (PAL; EC: 4.3.1.24 U min^{-1} mg^{-1} $protein^{-1}$)						
Control	7.47 ± 0.269 ^{j}	8.95 ± 0.076 ^{h}	12.35 ± 0.513 ^{ef}	15.71 ± 0.473 ^{d}	16.95 ± 0.404 ^{c}	24.31 ± 0.629 ^{a}
GA/PVP–SA 0 mM	7.47 ± 0.269 ^{j}	8.55 ± 0.026 ^{hi}	10.49 ± 0.348 ^{g}	12.86 ± 0.566 ^{e}	15.31 ± 0.294 ^{d}	18.50 ± 0.440 ^{b}
GA/PVP–SA 1 mM	7.47 ± 0.269 ^{j}	8.03 ± 0.030 ^{ij}	8.92 ± 0.092 ^{h}	10.58 ± 0.461 ^{g}	11.86 ± 0.263 ^{f}	12.68 ± 0.556 ^{e}
GA/PVP–SA 2 mM	7.47 ± 0.269 ^{j}	7.85 ± 0.097 ^{ij}	8.29 ± 0.023 ^{hi}	9.03 ± 0.040 ^{h}	9.85 ± 0.156 ^{g}	10.22 ± 0.190 ^{g}
Polyphenol oxidase activity (PPO; EC: 1.14.18.1 U min^{-1} mg^{-1} $protein^{-1}$)						
Control	0.26 ± 0.005 ^{n}	0.34 ± 0.008 ^{j}	0.44 ± 0.008 ^{f}	0.52 ± 0.012 ^{d}	0.58 ± 0.008 ^{b}	0.65 ± 0.011 ^{a}
GA/PVP–SA 0 mM	0.26 ± 0.005 ^{n}	0.31 ± 0.005 ^{l}	0.39 ± 0.005 ^{h}	0.46 ± 0.005 ^{e}	0.54 ± 0.005 ^{c}	0.59 ± 0.005 ^{b}
GA/PVP–SA 1 mM	0.26 ± 0.005 ^{n}	0.29 ± 0.005 ^{m}	0.33 ± 0.005 ^{k}	0.37 ± 0.005 ^{i}	0.40 ± 0.005 ^{h}	0.43 ± 0.005 ^{g}
GA/PVP–SA 2 mM	0.26 ± 0.005 ^{n}	0.26 ± 0.003 ^{n}	0.29 ± 0.005 ^{m}	0.31 ± 0.005 ^{l}	0.33 ± 0.005 ^{k}	0.35 ± 0.005 ^{j}

The interaction between GA/PVP–SA coating treatments and shelf-life duration in days when both are considered as factors. Means in the same column that have different letter(s) are significantly different using Duncan's multiple range test at $p < 0.05$ and ±SE (n = 3) replicates.

The results showed that the activities of PPO and PAL increased in both uncoated and coated peach fruits throughout the storage period (Table 2). Nevertheless, PPO and PAL activities in coated fruits increased slowly in comparison with control fruits. After 10 days of the experiment, minimum activities of PPO and PAL were noted in fruits dipped in GA/PVP–SA 2 mM (0.35 and 10.22 U min^{-1} mg $protein^{-1}$), while the highest activities of PPO and PAL were recorded in uncoated fruits (0.65 and 24.31 U min^{-1} mg $protein^{-1}$, Table 2, respectively).

2.4. The Activities of Cell-Wall-Degrading Enzymes (CWDEs)

Table 3 shows the alteration in the activities of CWDEs, i.e., LOX, CEL, and PG (U min^{-1} mg $protein^{-1}$), throughout the shelf-life period (days) for 'Early sweet' peach fruit. A significant interaction at $p \leq 0.001$ was observed when coating treatment of GA/PVP–SA and storage periods (days) were considered as a factor. Noticeably, all the activities of CWDEs increased gradually up to the sixth day of the experiment. Moreover, both LOX and PG enzymes continued to increase in activity until the 10th day of postharvest life; however, the CEL activity decreased. As presented in Table 3, the control treatment presents the highest significant activities of LOX (2.28 U min^{-1} mg $protein^{-1}$), CEL (17.85 U min^{-1} mg $protein^{-1}$), and PG (2.86 U min^{-1} mg $protein^{-1}$) on the 10th day. Conversely, the GA/PVP–SA 2 mM treatment presents the highest significant reduction in CWDEs, i.e., LOX (0.99 U min^{-1} mg $protein^{-1}$), CEL (9.64 U min^{-1} mg $protein^{-1}$), and PG (0.43 U min^{-1} mg $protein^{-1}$), at the end of the experiment.

Table 3. Effect of different treatments of gum arabic (GA) and polyvinylpyrrolidone (PVP) blended with salicylic acid (SA) at different concentrations as a mixture biopolymer coating treatment (GA/PVP–SA) on cell-wall-degrading enzyme activities of peach fruit 'Early Sweet' during shelf life at ambient air (29 ± 1 °C and air humidity 51%) during two seasons (2020 and 2021).

Treatments	Shelf-Life Period (Days)					
	D_0	D_2	D_4	D_6	D_8	D_{10}
Lipoxygenase activity (LOX; EC:1.13.11, U min^{-1} mg^{-1} $protein^{-1}$)						
Control	0.84 ± 0.008 ^{m}	0.93 ± 0.023 ^{ij}	0.96 ± 0.021 ^{hi}	1.31 ± 0.020 ^{e}	1.94 ± 0.032 ^{b}	2.28 ± 0.015 ^{a}
GA/PVP–SA 0 mM	0.84 ± 0.008 ^{m}	0.88 ± 0.003 ^{l}	0.92 ± 0.005 ^{jk}	1.00 ± 0.012 ^{g}	1.73 ± 0.043 ^{c}	1.94 ± 0.023 ^{b}
GA/PVP–SA 1 mM	0.84 ± 0.008 ^{m}	0.86 ± 0.008 ^{lm}	0.88 ± 0.006 ^{kl}	0.94 ± 0.005 ^{ij}	1.11 ± 0.017 ^{f}	1.35 ± 0.017 ^{d}
GA/PVP–SA 2 mM	0.84 ± 0.008 ^{m}	0.84 ± 0.005 ^{m}	0.86 ± 0.003 ^{lm}	0.89 ± 0.005 ^{kl}	0.93 ± 0.008 ^{ij}	0.99 ± 0.005 ^{gh}
Cellulase activity (CEL; EC: 3.2.1.4 U min^{-1} mg^{-1} $protein^{-1}$)						
Control	8.00 ± 0.063 ^{i}	13.03 ± 0.777 ^{f}	15.42 ± 0.666 ^{e}	22.67 ± 1.353 ^{a}	19.65 ± 0.619 ^{b}	17.85 ± 0.543 ^{c}
GA/PVP–SA 0 mM	8.00 ± 0.063 ^{i}	10.65 ± 0.450 ^{g}	12.66 ± 0.538 ^{f}	16.67 ± 0.390 ^{d}	14.76 ± 0.459 ^{e}	12.72 ± 0.592 ^{f}
GA/PVP–SA 1 mM	8.00 ± 0.063 ^{i}	9.61 ± 0.363 ^{h}	10.76 ± 0.453 ^{g}	14.71 ± 0.509 ^{e}	12.43 ± 0.738 ^{f}	10.96 ± 0.383 ^{g}
GA/PVP–SA 2 mM	8.00 ± 0.063 ^{i}	8.50 ± 0.100 ^{i}	9.52 ± 0.272 ^{h}	11.00 ± 0.455 ^{g}	10.41 ± 0.057 ^{gh}	9.64 ± 0.048 ^{h}
Pectinase activity (PT; EC: 3.2.1.15, U min^{-1} mg^{-1} $protein^{-1}$)						
Control	0.27 ± 0.005 ^{o}	0.36 ± 0.005 ^{klm}	1.46 ± 0.014 ^{d}	1.57 ± 0.005 ^{c}	1.81 ± 0.033 ^{b}	2.86 ± 0.067 ^{a}
GA/PVP–SA 0 mM	0.27 ± 0.005 ^{o}	0.33 ± 0.005 ^{mn}	0.40 ± 0.003 ^{jk}	0.50 ± 0.005 ^{i}	1.03 ± 0.040 ^{f}	1.22 ± 0.063 ^{e}
GA/PVP–SA 1 mM	0.27 ± 0.005 ^{o}	0.30 ± 0.003 ^{no}	0.38 ± 0.003 ^{jkl}	0.44 ± 0.005 ^{j}	0.64 ± 0.008 ^{h}	0.72 ± 0.005 ^{g}
GA/PVP–SA 2 mM	0.27 ± 0.005 ^{o}	0.28 ± 0.003 ^{no}	0.34 ± 0.005 ^{lmn}	0.39 ± 0.006 ^{jk}	0.42 ± 0.005 ^{j}	0.43 ± 0.005 ^{j}

The interaction between GA/PVP–SA coating treatments and shelf-life duration in days when both are considered factors. Means in the same column that have different letter(s) are significantly different using Duncan's multiple range test at $p < 0.05$ and ± SE (n = 3) replicates.

2.5. Cell Membrane: Lipid Peroxidation (MDA; $\mu M\ mg^{-1}$ FW) and Electrolyte Leakage (EL)

Experimental data of MDA and EL% as indicators of membrane disruption are presented in this study. Table 4 exhibits a significant interaction at $p < 0.001$ between postharvest time (days) and GA/PVP–SA treatments. Perceptibly, both cell membrane termination and MDA and EL% increased gradually over all treatments up to the 10th day of the postharvest period, contrasting with the initial rates at harvest time. Moreover, the control treatment showed more rapid MDA accumulation and ion permeability percentage during the postharvest period compared to other coating treatments. The coating treatment of GA/PVP–SA 2 mM is recognized as the most valuable one in minimizing the alternations in cell membrane parameters, where it recorded the values of 0.32 μM g^{-1} FW and 13.24% for MDA and EL%, respectively, on the 10th day of storage time.

Table 4. Effect of different treatments of gum arabic (GA) and polyvinylpyrrolidone (PVP) blended with salicylic acid (SA) at different concentrations as a mixture biopolymer coating treatment (GA/PVP–SA) on lipid peroxidation (MDA) and cell membrane leakage percentage of peach fruit 'Early Sweet' during shelf life at ambient air (29 ± 1 °C and air humidity 51%) during two seasons (2020 and 2021).

Treatments	Shelf-Life Period (Days)					
	D_0	D_2	D_4	D_6	D_8	D_{10}
Lipid peroxidation (MDA; μM g^{-1} FW)						
Control	0.19 ± 0.005 ^{n}	0.29 ± 0.012 ^{j}	0.45 ± 0.008 ^{e}	0.56 ± 0.020 ^{c}	0.59 ± 0.011 ^{b}	0.68 ± 0.014 ^{a}
GA/PVP–SA 0 mM	0.19 ± 0.005 ^{n}	0.26 ± 0.005 ^{k}	0.46 ± 0.005 ^{e}	0.40 ± 0.005 ^{f}	0.46 ± 0.005 ^{e}	0.50 ± 0.005 ^{d}
GA/PVP–SA 1 mM	0.19 ± 0.005 ^{n}	0.24 ± 0.005 ^{l}	0.29 ± 0.005 ^{j}	0.31 ± 0.012 ^{ij}	0.37 ± 0.005 ^{g}	0.39 ± 0.005 ^{f}
GA/PVP–SA 2 mM	0.19 ± 0.005 ^{n}	0.21 ± 0.005 ^{m}	0.22 ± 0.005 ^{m}	0.25 ± 0.005 ^{kl}	0.30 ± 0.005 ^{j}	0.32 ± 0.005 ^{i}
Electrolyte leakage (EL%)						
Control	6.67 ± 0.273 ^{p}	12.45 ± 0.631 ^{kl}	18.75± 0.548 ^{g}	24.89 ± 1.132 ^{e}	35.20 ± 0.273 ^{b}	52.88 ± 1.475 ^{a}
GA/PVP–SA 0 mM	6.67 ± 0.273 ^{p}	11.09± 0.147 ^{mn}	15.91 ± 0.419 ^{i}	18.79 ± 0.275 ^{g}	26.28 ± 0.273 ^{d}	30.44 ± 0.419 ^{c}
GA/PVP–SA 1 mM	6.67 ± 0.273 ^{p}	10.08 ± 0.037 ^{n}	14.34 ± 0.328 ^{j}	17.13 ± 0.273 ^{h}	20.63 ± 0.273 ^{f}	26.76 ± 0.539 ^{d}
GA/PVP–SA 2 mM	6.67 ± 0.273 ^{p}	7.59 ± 0.117 ^{p}	8.86 ± 0.150 ^{o}	10.45± 0.273 ^{mn}	11.62± 0.273 ^{lm}	13.24± 0.541 ^{jk}

The interaction between GA/PVP–SA coating treatments and shelf-life duration in days when both are considered factors. Means in the same column that have the different letter(s) are significantly different using Duncan's multiple range test at ($p < 0.05$) and ±SE (n = 3) replicates.

2.6. Fruit Ethylene and Respiration

Figure 1 shows the differences in ethylene production and respiration rates of peaches during storage in shelf-life conditions. Each of the two gases rose to a maximum on the second day that is higher than the initial value. During the shelf-life continuance, ethylene increases clearly to a maximum of three times and, for the respiration, 2.5 times on the second day for the control treatment. The increases are independent according to the treatments. GA/PVP–SA 2 mM presented more inhibition in both ethylene and respiration throughout the experiment (Table 5). It recorded 11.17 and 14.57 mg kg^{-1} h^{-1} for both gasses on the second day of the shelf-life period. Consequently, it minimizes both ethylene and carbon dioxide production until up to the end of the storage time (Figure 2).

Table 5. Effect of different treatments of gum arabic (GA) and polyvinylpyrrolidone (PVP) blended with salicylic acid (SA) at different concentrations as a mixture biopolymer coating treatment (GA/PVP–SA) on ethylene and respiration production rate of peach fruit 'Early Sweet' during shelf life at ambient air (29 ± 1 °C and air humidity 51%) during two seasons (2020 and 2021).

Treatments	Shelf-Life Period (Days)					
	D_0	D_2	D_4	D_6	D_8	D_{10}
Ethylene concentration (C_2H_4; mg kg^{-1} h^{-1} FW)						
Control	5.71 ± 0.340 ^{kl}	24.64 ± 0.583 ^{a}	22.25 ± 0571 ^{b}	18.52 ± 0.586 ^{c}	14.76 ± 0.583 ^{de}	11.16 ± 0.554 ^{gh}
GA/PVP–SA 0 mM	5.71 ± 0.340 ^{kl}	22.43 ± 0.571 ^{b}	19.17 ± 0.591 ^{c}	15.14 ± 0.318 ^{d}	12.25 ± 0.554 ^{fg}	9.23 ± 0.560 ^{i}
GA/PVP–SA 1 mM	5.71 ± 0.340 ^{kl}	17.73 ± 0.574 ^{c}	13.32 ± 0.886 ^{ef}	11.32 ± 0.571 ^{gh}	9.38 ± 0.562 ^{i}	7.21 ± 0.333 ^{jk}
GA/PVP–SA 2 mM	5.71 ± 0.340 ^{kl}	11.17 ± 0.606 ^{gh}	9.76 ± 0.307 ^{hi}	8.23 ± 0.571 ^{ij}	7.28 ± 0.355 ^{jk}	5.00 ± 0.330 ^{l}
Respiration evolution (CO_2; mg kg^{-1} h^{-1} FW)						
Control	12.91 ± 0.597 ^{fg}	33.16 ± 0.586 ^{a}	32.26 ± 0.560 ^{a}	28.17 ± 0.640 ^{b}	24.03 ± 0.331 ^{c}	19.16 ± 0.552 ^{e}
GA/PVP–SA 0 mM	12.91 ± 0.597 ^{fg}	27.32 ± 0.568 ^{b}	23.82 ± 0.869 ^{cd}	22.17 ± 0.346 ^{d}	19.23 ± 0.568 ^{e}	14.62 ± 0.586 ^{f}
GA/PVP–SA 1 mM	12.91 ± 0.597 ^{fg}	22.42 ± 0.577 ^{cd}	18.41 ± 0.902 ^{e}	14.63 ± 0.853 ^{f}	13.36 ± 0.586 ^{f}	9.76 ± 0.336 ^{hi}
GA/PVP–SA 2 mM	12.91 ± 0.597 ^{fg}	14.54 ± 0.588 ^{f}	11.14 ± 0.586 ^{gh}	10.52 ± 0.574 ^{hi}	8.28 ± 0.353 ^{ij}	7.83 ± 0.568 ^{j}

The interaction between GA/PVP–SA coating treatments and shelf-life duration in days when both are considered factors. Means in the same column that have different letter(s) are significantly different using Duncan's multiple range test at $p < 0.05$ and ±SE (n = 3) replicates.

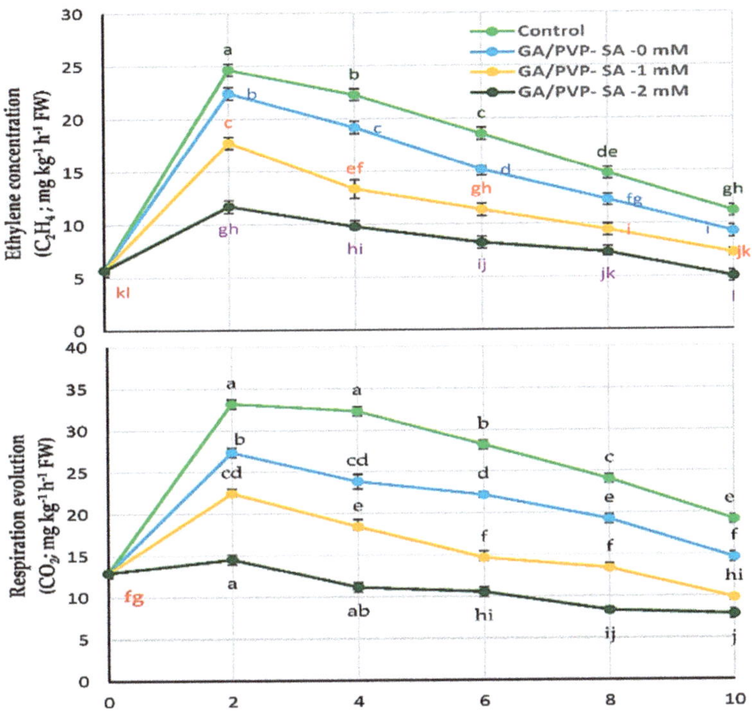

Figure 2. Effect of different treatments of gum arabic (GA) and polyvinylpyrrolidone (PVP) blended with salicylic acid (SA) at different concentrations as a mixture biopolymer coating treatment (GA/PVP–SA) on ethylene and respiration production rate of peach fruit 'Early Sweet' during shelf life at ambient air (29 ± 1 °C and air humidity 51%) during two seasons (2020 and 2021).

2.7. Data Modeling

Linear regression analysis was performed to predict the properties of the treated fruit with the best treatment (AG/PVP–SA 2 mM). It can be concluded from Table 6 that R^2 values of WL%, firmness, MDA, EL%, and the enzyme activities of PAL, PPO, and PT are greater than 0.90, meaning that the linear model could be properly used for forecasting the characteristics of the treated peach with AG/PVP–SA 2 mM at different storage periods from 0 to 10 days. Meanwhile, the linear model cannot be useful for predicting ethylene, SBI, and CEL due to its lower R^2 values, which were less than 0.7.

Table 6. Modeling of changes in the properties during the shelf life of the treated peach with GA/PVP–SA 2 mM.

Properties	Linear Model ($Y = a \pm bX$) *		
	a (p Value)	b (p Value)	R^2
C_2H_4	8.99 (0.00)	0.224 (0.23)	0.101
CO_2	14.11(0.00)	−0.64 (0.00)	0.753
WL%	0.287 (0.231)	1.42 (0.00)	0.988
SBI	0.997 (0.00)	0.0009 (0.015)	0.357
Hue	8.28 (0.00)	1.19 (0.00)	0.814
Firmness	47.86(0.00)	−0.82(0.00)	0.901

Table 6. *Cont.*

Properties	Linear Model ($Y = a \pm bX$) *		
	a (p Value)	b (p Value)	R^2
TP	88.00(0.00)	−0.82(0.00)	0.755
TF	34.61(0.00)	−0.95(0.00)	0.892
PAL	7.43(0.00)	0.27(0.00)	0.949
PPO	0.26(0.00)	0.01(0.00)	0.933
LOX	0.82(0.00)	0.01(0.00)	0.865
CEL	8.35 (0.00)	0.242(0.002)	0.512
PT	0.27(0.00)	0.02(0.00)	0.931
MDA	0.18(0.00)	0.01(0.00)	0.929
EL%	6.53(0.00)	0.61(0.00)	0.949

* Y and X denote the dependent (properties) and independent (shelf-life duration) variables, respectively.

3. Discussion

The IR spectrum of salicylic acid (Figure 1) reveals the absorption bands ascribed to the stretching vibrations of O–H bonds of phenyl hydroxyls (3238 cm^{-1}), and the stretching (a shoulder at 3063 cm^{-1}) and bending (1297 cm^{-1}) vibrations of C–H bonds of aromatic rings, C=O bonds (1659 cm^{-1}), and C=C bonds of benzene rings (1612, 1577, 1483, and 1443 cm^{-1}). Overall, the FT-IR spectrum of SA in this study agreed to a large extent with that obtained by [40].

For GA, the broad peak observed at 3422 cm^{-1} is ascribed to OH groups of the carbohydrate structure. Meanwhile, the major IR bands observed at 2927 cm^{-1} were assigned to the vibrational modes of C–H groups. Peaks present in the spectra at 1654 and 1429 cm^{-1} correspond to the occurrence of the carboxylic groups. It is well known that carboxylic acids display a characteristic OH in-plane bending band at 1430 cm^{-1} [41]. Therefore, the peak found at 1429 cm^{-1} wavenumber may be due to the symmetrical stretching of uronic acid carboxylates in the structure of GA. The peaks found between 800 cm^{-1} and 1200 cm^{-1} represented C–C, C–O, and C–O–C stretching and C-OH and C–H bending modes of the polymer backbone. Bands detected in the spectra at 776 cm^{-1} may be assigned to the 1–4 linkage of galactose and 1–6 linkage of mannose [41]. The peaks observed at 1030 and 879 cm^{-1} in the FT-IR curve of GA may be assigned to arabinogalactan. In the present work, the bands found between 700 and 500 cm^{-1} were attributed to the pyranose rings. Our results concerning the FT-IR spectrum of GA were in accordance with those obtained by [41].

For PVP, the peaks observed at 3450 cm^{-1} are assigned to hydroxyl stretching, and the peaks at 1435 and 845 cm^{-1} correspond to the CH_2 scissoring vibrations and CH_2 bending, respectively. The peaks at 1656 and 1289 cm^{-1} are assigned to C=O stretching and C–N stretching. Our results concerning the FT-IR spectrum of PVP were in accordance with those obtained in [42].

In the GA/PVP blend, the C–N bending vibration from the PVP pyrrolidone structure was shifted to be at 1294 cm^{-1}, while the peak at 776 cm^{-1} in this blend confirmed the presence of 1–4 linkage of galactose and 1–6 linkage of mannose related to GA. The FT-IR spectrum of GA/PVP–SA 2 mM has prominent bands at 3240, 1660, 1612, and 1483 cm^{-1} wavenumbers due to the presence of SA in the polymer blend. Meanwhile, the peak found at 1296 cm^{-1} might be related to the pyrrolidone structure of PVP. The board peak of hydroxyl groups related to GA shifts to the wavenumber of 3528 cm^{-1}. Moreover, the shifted peak at 760 cm^{-1} related to the 1–4 linkage of galactose and 1–6 linkage of mannose might support the presence of GA in GA/PVP–SA 2 mM treatment.

Coating polymers are generally made of fats, proteins, and polysaccharides that inhibit water loss [15,43] and probably remain less prone to microbial attack [44,45]. The reason for decreasing water loss by GA/PVP blends in this study is principally due to the properties of GA arabinogalactan (80–90%) in the retaining of water, which prevents the loss of fruits' water during the storage period [15]. Similarly, the authors of [6] found that two varieties of *Aloe* gels as a coating source were able to diminish water loss of palm and peach fruits compared with uncoated fruits. Polysaccharide biodegradable coatings can decrease water loss of the fruit tissues by forming a physical barrier around the fruit surface [41]. Overall, the efficiency of a polysaccharide polymer as a biodegradable coating critically depends on its physical properties [42]. It is thought that high molecular weight polysaccharides, such as GA, have huge mechanical properties [42]. PVP is a synthetic polymer that forms a hydrogel that can preserve a huge amount of water.

In addition, using PVP increases coating performance to maintain peach fruit quality and decline the water evaporation. Thus, the use of pure PVP is restricted. To overcome this difficulty, PVP is fabricated with GA in this study to obtain a blend possessing acceptable properties, i.e., forming a physical barrier and preserving a high amount of water [46].

The presence of phenolic ingredients increases cellular immunity due to their biological properties, such as antioxidant capacity and protection against fungi and bacteria [43]. GA/PVP–SA 2 mM presents a more effective treatment that maintains the phenolic load throughout the shelf-life period of peach fruits. Our findings agreed to a large extent with those obtained by [44], who found that 1.5 mM SA-treated peach fruits kept the higher significant amounts of ascorbic acid, TFs, and TPs and antioxidant capacity when compared to untreated fruits.

Enzymatic browning is a common phenomenon that can usually be detected in different fruits, which unfavorably affects the nutritional value and other quality attributes. This happens when the phenolic ingredients are oxidized by PPO to their quinone derivatives and, additionally, are oxidized to form melanin pigment accountable for the browning reactions [45]. In the present study, GA/PVP–SA coatings can decrease the incidences of skin browning in peach fruit by decreasing PPO and PAL activities. The effectiveness of GA/PVP coatings in decreasing the action of browning enzymes could be due to their supplementation with salicylic acid. In this respect, the authors of [31] stated that SA shows an important role in the inhibition of the activities of browning-related enzymes in guava fruits.

The impact of SA on phenolic contents and the activities of browning enzymes depend on some critical factors, such as type of stress, SA concentration, availability, plant variety, postharvest conditions, and so on. In this respect, postharvest treatment of peach fruit with SA at 2 mM recorded lower PPO activity, accompanied by higher activities of antioxidant enzymes during shelf-life storage [46]. Moreover, another study indicated that salicylic acid alleviated chilling incidence, reduced PAL activity, and preserved phenolics and antioxidant capacity in pomegranate fruit in the postharvest period [47]. Showing different behavior, hot salicylic acid preserved higher anthocyanin and total phenolic contents in the arils of pomegranate during postharvest storage at 4 °C for nearly two weeks via diminishing PPO activity in combination with better PAL activity [7,48,49]. Similarly, improved total phenol content in SA- and calcium-chloride-treated cornelian cherry fruits may be attributed to higher PAL activity [6]. On the contrary, the application of SA in sponge gourds significantly decreased TPs due to its ability to inhibit the activities of PAL and PPO browning enzymes [50,51]. In our study, however, SA incorporated into the GA/PVP coating can decrease the activities of PPO and PAL browning enzymes. The increased amounts of polyphenols in GA/PVP–SA-treated fruits might be due to the higher PAL/PPO ratio. Overall, the higher accumulation of phenolic compounds in fruit tissues without undesirable browning incidences is largely correlated to a higher PAL/PPO ratio [48].

Skin browning is a common problem for extending the postharvest life of peach fruit, which is most sensitive to mechanical injury. Browning is chiefly caused by the enzymatic oxidation of endogenous phenols into quinones [47]. SBI weighs the clarity of the brown color and is reflected as a critical factor for examining the types of browning in fruits [52]. It was detected that SBI in the control peach fruits had extensive variations and specifically increased after 6 days of storage. This finding agreed with the obtained highest level of PPO activity in the control group throughout the storage period (Table 1). Meanwhile, the coated fruits did not display any significant alteration over the whole period of storage, reflecting the ability of GA/PVP blends to prevent browning incidences as a result of their ability to decrease the activity of browning enzymes.

The acceptable color and appearance of peach fruit is the critical factor for its friendly marketing. The hue angle of vegetables and fruits was mostly affected by the coating treatment. Moreover, coated peach fruits' color should stay parallel to a fresh one by combating any deterioration of color during storage. In this regard, the little rate of decrease in hue angle of peach fruits coated with GA/PVP blends in comparison to uncoated fruits reflects the effectiveness of these polymer composites in avoiding color rapid deteriorations. Our results agreed with [7], who showed that peach fruits coated with *Aloe vera* gel under shelf-life storage over 30 days had a rate of diminished hue angle that was much slighter than the uncoated fruits. It could be suggested that GA/PVP blends can prevent color rapid deterioration [48]. The proper role of SA in polymer blends in inhibiting the enzymes that cause tissue softness may reflect the retardation of color changes in peach fruit treated with GA/PVP–SA 2 mM.

During the ripening process, fruit firmness decreased gradually due to the induction of the activities of CWDEs. In this study, GA/PVP formulations could preserve peach fruit firmness, especially GA/PVP–SA 2 mM. The literature data revealed a discrepancy in the impact of different edible coatings on peach fruit firmness, validating the proper role of SA in the present study. For instance, it was found that *Aloe* gels had no effect on the level of firmness in peach fruit stored at 20 °C for six days; its values gradually decreased in coated and uncoated fruits alike, with no significance [49]. In contrast, a significant impact beginning from the 10th day of *A. vera* coating film decreasing the firmness loss of peach fruits stored at 4 °C has been recently reported [7].

The present data elucidated the efficient role of GA/PVP biopolymer coatings in diminishing the activities of PG, CEL, and LOX (Table 3). Thus, the softness retardation of peach fruits treated by GA/PVP coatings in this study may be due to the property of GA and SA alone to inhibit the activities of CWDEs, which preserve firmness for as long as possible. In this respect, the inhibitory effect of GA/chitosan (10:1) composite against the activities of CWDEs, and thus keeping the level of firmness in stored banana fruit for as long as possible, has been documented [15]. Meanwhile, the ability of chitosan-based polymer blends to suppress the activities of CWDEs in guava fruits was significantly increased in the presence of SA, thereby reducing tissue breakdown and fruit water loss and preserving the level of firmness in three phases of fruit maturity [31].

The plant cell wall is a complex reticulate structure, which consists of structural proteins, pectin, cellulose, and hemicellulose [50]. Pectin is the chief element in the cell's primary wall and the middle lamella and can tie cells together, similarly to 'glue'. During ripening, fruit firmness decreases regularly due to the initiation of the activities of CWDEs, such as PG, LOX, and CEL. Many applications have been stated to reduce ethylene production and, thereby, prevent the activity of the enzymes, delaying softness.

The hydrolysis of pectin is catalyzed by related enzymes comprising pectin methylesterases, PG, pectate-lyase, and β-galactosidase, of which PG has been proposed to act as a vital role [51]. Moreover, gene expression of PGs during softening of two peach fruit cultivars with different softening features has been recently identified [52]. Thus, in this study, the activity of PG was examined as an appropriate indicator of cell wall pectin hydrolysis. The maximum obtained level of PG activity in control uncoated fruits in this study clarified their rapid ripening and softening.

Cellulases (E.C. 3.2.1.4) hydrolyze β-1,4 linkages of cellulose, cellobiose, and cellodextrin. Generally, they are multienzyme complexes having endo-1,4-β-glucanase, β-glucosidase, and cellobiohydrolase activity [53]. In fruits, cellulase activity is mostly correlated to softening physiological processes during maturation. The highest obtained level of CEL activity in control untreated fruits in this study elucidated their rapid ripening and loss of firmness. A high correlation between the huge level of cellulase activity and the minimum level of firmness in fruit tissues has been stated [54]. Enzymatic depolymerization of hemicellulose plays a key role in fruit maturation, leading to the disassembly of hemicellulose and the cellulose network and a reduction in fruit firmness [55]. Based on the obtained results, the increase in CEL activity in coated and uncoated fruits until the 6th day may be due to the disassembly of the hemicellulose/cellulose network. The decline in CEL activity after the 6th day in all treatments could be due to the extensive alternation in the hemicellulose structure.

In this study, the GA/PVP biopolymer coating efficiently diminished the activities of PG and CEL. These desired effects progressively increased in the presence of salicylic acid in the polymer composites. Inhibition of fruit CWDEs and preservation of firmness using edible coating formulations have been noted by several researchers. In this respect, Dave et al. [56] found that the formulations depend on hydroxyl-propyl methylcellulose; soy protein isolate and olive oil had an inhibitory effect on the activities of enzymes related to fruit softening, including pectin methylesterase, β-galactosidase, and PG, in pears stored at 28 °C. Srivastava and Dwivedi [44] found that 1 mM SA was able to delay softening by decreasing the activities of PG, CEL, and xylanase in bananas. Moreover, supplementation of chitosan-based polymer blends with SA has been reported to decline the activities of CWDEs in guava fruit [31]. Overall, earlier findings explained the role of SA application in the enhancement of the activity of polymer coatings.

Lipoxygenase (LOX, EC 1.13.11.12) is an enzyme that commonly exists in plant tissues, which activates the oxidation of polyunsaturated fatty acids to form corresponding hydroperoxides. The fatty acid hydroperoxides formed by the activity of LOX are possibly harmful to membrane function by initiating increased rigidity [47]. It also performs positively through its function in the development of defense-related signaling molecules [57]. Lipoxygenases possess some applications in food technology, such as aroma production and bread making; they also have undesirable effects, including off-flavor and color changes in different foods [47]. In this study, GA/PVP/SA biopolymer coatings effectively diminished the activity of LOX. This desired effect could be due to the presence of salicylic acid in the polymer composite. In this respect, the authors of [47] found a reduction in the expression of *DkLOX*$_3$ by SA, concomitant with the preservation of fruit firmness, inhibition of weight loss, and ethylene production during persimmon fruit storage. Moreover, the inhibitory effect of nitric oxide on LOX and ethylene biosynthesis in the shelf life of peach fruit has been previously reported [58]. The highest level of LOX activity of uncoated peach fruits at the end of the storage period in this study agreed with the obtained results regarding water loss and softness. Overall, the activity of LOX and other lipolytic enzymes increases during senescence [59], the earlier activity causing the leakage of membrane polyunsaturated fatty acids (PUFAs) that can act as a substrate for LOX. In response to wounding or senescence, LOX could be complicated undesirably through contribution to autocatalytic peroxidation reactions [60]. The resultant hydroperoxides can induce tissue injury through protein synthesis inactivation and dysfunction of cellular membranes. Lastly, the inhibitory effect of GA/PVP/SA biopolymer coatings on lipolytic enzymes in this study will offer new indications for exploring the roles of *LOX* in delaying peach fruit ripening and preserving firmness.

MDA is the product of lipid peroxidation, and its accumulation is indicative of cell membrane degradation. It was previously described that the increased amount of MDA is principally due to the increased activity of LOX [61]. Our results concerning lipid peroxidation and ion permeability suggest that GA/PVP–SA is a promising tool for avoiding postharvest oxidative damage.

The usage of biodegradable coatings alone or incorporated with bioactive additives in fruit postharvest technology significantly reduces the accumulation rate of MDA, which preserves the functions of cellular membranes and, thus, also reduces cell permeability rates [62]. This could be interpreted by the facilitation of coating material creating a barrier to the oxygen responsible for lipid peroxidation, hence maintaining membrane integrity [62]. Besides, GA was previously reported to delay ripening and consecutively preserve the antioxidant status of tomatoes up to three weeks after harvest [14]. Furthermore, the presence of SA in chitosan-based polymer blends may play a vital role in the inhibition of tissue breakdown by inhibiting the activities of CWDEs [31,57], as well as ethylene gas production and respiration [22]. Salicylic acid was also stated to reduce fruit senescence during shelf life [1,36].

The respiratory performance of GA/PVP/SA-coated peach fruits (Table 5) presented delayed the attainment of respiratory climacteric, and the respiratory magnitude was also found to be significantly ($p < 0.05$) lower compared to uncoated fruits after the second day until the end of the storage period. Polymer coatings acted as a barrier film, providing a different internal atmosphere and a selective membrane for permeation of ethylene in and out of the fruit, as well as diminished production of ethylene by the fruit tissues [31]. Moreover, the inhibitory action of SA on the conversion of 1-aminocyclopropane-1-carboxylic acid (ACC) into ethylene via decreasing the activity of ACC oxidase has been reported [63].

The influence of SA on ethylene production was studied on different fruits, i.e., tomato [64], apple [65], and Selva strawberry fruit [63]. Lastly, a coating treatment of GA/PVP–SA 2 mM effectively decreased ethylene production in peach fruit, which might be explained by the role of SA in minimizing the respiration rate by increasing the energy charge [38].

4. Materials and Methods

4.1. Fruit Material

The present study was conducted throughout the season (2020 and 2021) on peach fruit (*Prunus persica*) 'Early Sweet' cv. Trees were planted in a commercial orchard in Meet-Gamer province, Egypt, in clay soil that is 11 years old. All trees were grafted on 'Nemaguard' rootstock at 4×4 m^2 plantation distance. The trees were also pruned in an open vase shape under drip irrigation. The ranch administration program was connected by the proposals of the Egyptian Agricultural Ministry. Fruits were selected from the shaded side of trees in the same uniform at full maturation in May [66]. At harvest time, the fruit was selected when fruits reached SSC 8.5% and fruit firmness 47 N. A total of 360 fruits were harvested and divided into two main lots. The first lot contained 180 fruits dedicated to non-distractive measurements. Every treatment contained 45 fruits (15 fruits per replicate).

4.2. Gum Arabic and Coating Protocol

Gum arabic (GA) powder was supplied by El-Gomhoria Company, Cairo city, Egypt. It was prepared by adding 60 g GA to 500 mL distilled water. Thereafter, the solution was stirred by hot magnetic stirring at 55 °C for 60 min. Then, the solution was cooled at room temperature and filtered to remove impurities using a fine piece of cloth [14]. To prepare PVP (K-20 polymer, Ashland company, Shanghai, China), 15 g was dissolved in 500 mL of distilled water under magnetic stirring until dissolved. The polyvinylpyrrolidone (PVP) was supplemented with the GA solution in a ratio of (1:1) to form a coating mixture with desirable mechanical and physical properties. The resultant GA/PVP blend was then stirred for one hour. The final concentrations of GA and PVP in the resultant blend were 6% and 1.5%, respectively. To increase the covering mixture effectiveness, SA was supplemented to the mixture at three concentrations (0, 1, and 2 mM). Finally, the blended mixture solution was prepared in three main stocks GA/PVP–SA (0 mM), GA/PVP–SA (1 mM), and GA/PVP–SA (2 mM), each in a volume of 2 L. Afterward, fruits were soaked in the prepared mixtures and stored at 24 ± 3 °C and air humidity at 51 ± 1%.

4.3. Fourier Transform Infrared Analysis (FT-IR)

For identification of the functional groups and chemical bonding in pure SA, GA, PVP, and the dry powders of blend solutions of GA/PVP and GA/PVP-SA 2 mM, 1 mg of each sample was mixed with 300 mg of fine potassium bromide (KBr). The thin pellets were prepared by pressing with the hydraulic pellet press and were then subjected to

Fourier transform infrared spectrophotometer (FTIR, Jinan City, China) in the range of 500–4000 cm^{-1} at a resolution of 4 cm^{-1} [40–42].

4.4. Water Loss %, Skin Browning Index, and Fruit Color Hue Angle

Peach fruit samples (15 fruit per replicate) were weighed at harvest time up to the end of the experiment time interval (2 days). Weight loss % was computed based on the initial value at harvest time [31].

As for fruit skin, the browning parameter was determined by browning spots visually during shelf-life duration. The browning scale was identified in five categories [67]. The categories are: 1 = no brown spots; 2 = slight browning; 3 = moderate browning spot; 4 = severe browning symptoms; and 5 = very severe symptoms. Meanwhile, the fruit color hue angle parameter was assessed according to the RGB protocol [9].

Fruit firmness was measured as the necessary force to penetrate the tissue using the fruit texture Effegi penetrometer (Effegi, 48,011 Alfonsine, Alfonsine, Italy) [33].

4.5. The Browning Parameters: Total Phenols (TPs) and Total Flavonoids (TFs)

TPs content of peach fruit samples was evaluated using Folin–Ciocalteu reagent and gallic acid as a standard phenolic compound. TPs content was spectrophotometrically estimated at 750 nm. Data were represented as mg 100 g^{-1} fresh weight (FW) as gallic acid equivalents (GAE) [34]. Meanwhile, the flavonoid content was recorded in time intervals throughout the experimental time, and the data are shown as mg 100 g^{-1} FW catechin equivalent (CE) [35].

4.6. Extraction of Browning Enzyme

A gram of fruit sample in time was homogenized with 5 mL of Tris-HCl buffer (pH 7.0; 20 mM). Then, the mixture was centrifuged at 4 °C (15,000 rpm, 5 min) and the resultant supernatant was kept at −20 °C until further steps.

Polyphenol oxidase (PPO) (*EC: 1.14.18.1*) was assessed according to the procedure of [37] by mixing 0.5 mL catechol (500 mM) and 2 mL 0.05 M phosphate buffer, pH = 7.0, to 100 µL supernatant and incubated for 2 min at 24 °C. The enzyme activity was spectrophotometrically noted at 398 nm within 3 min. The enzyme activity was stated as units U min^{-1} mg $protein^{-1}$.

Phenylalanine ammonia-lyase (PAL) (*EC: 4.3.1.24*) activity was assessed spectrophotometrically at 290 nm according to the steps described by [37] after its extraction from one gram of the fruit sample with 4 mL of 200 mmol L^{-1} boric acid buffer (pH 8.8). The activity of PAL was lastly expressed as units U min^{-1} mg $protein^{-1}$.

4.7. The Activities of Cell-Wall-Degradation Enzymes (CWDEs) and Fruit Firmness

For enzyme extraction and assay, three grams of peach tissue was homogenized in 10 mL Tris-HCL (0.02 M, pH = 7.0) containing EDTA (20 mM), cysteine-HCL (0.02 M), and Triton X-100 (0.05%). Then, centrifugation was carried out at 15,000× *g* for half an hour at 4 °C. Lastly, the supernatant was kept at −20 °C until the enzymatic assessments [68]. The supernatant protein concentration was estimated according to Bradford's method [69].

The assay of polygalacturonase (PG) activity was evaluated according to the scheme designated by [39]. Volumes of 400 µL of Na-acetate buffer (0.2 M, pH 4.5), 600 µL polygalacturonic acid (PGA, 1%, pH 4.5), and 200 µL sodium chloride (0.2 M) were successively added to 100 µL of the supernatant, and the resultant mixture was maintained at 37 °C for 60 min. Then, 6 mL 3,5-dinitro salicylic acid (DNS) was added. The reaction was blocked by heating at 85 °C for 15 min. Finally, 1 mL of 40% sodium potassium tartrate was added. In control samples, the substrate was added after heating. The formed reducing groups were assessed against *D*-galacturonic acid after assessing the absorbance at 540 nm. PG activity was expressed as U min^{-1} mg $protein^{-1}$.

Assay of cellulase (CEL): the activity of CEL was evaluated according to the method described by [43]. The clear supernatant (250 µL) was added to the reaction mixture (500 µL carboxymethyl cellulose (1.5%) in 250 µL sodium acetate buffer (0.1 M, pH = 5.0)). The reaction mixture was incubated at 37 °C for 12 h. Then, 3 mL DNS was added, and the reaction was blocked by heating at 85 °C for 10 min. Finally, 1 mL of 40% SPT was added. In the control tube, the substrate was added after heating. The resultant reducing moieties were assessed against D-glucose after assessing the absorbance at 540 nm. CEL activity was expressed as U min^{-1} mg $protein^{-1}$.

Lipoxygenase (EC: 1.13.11; LOX) activity was evaluated relating to the scheme [43]. Lipoxygenase was extracted after homogenizing 2 g of peach fruit with 6 mL of extracting buffer (0.6 mL potassium phosphate buffer (0.5 M, pH = 7.8), 0.6 mL sodium-EDTA (0.01 M, pH = 7), and 4.8 mL of PVPP (2% *w*/*w*)). The mixture was then centrifuged for 20 min at 17,000 rpm at 4 °C. The resultant supernatant (0.1 mL) was mixed with 2.8 mL of sodium phosphate buffer (0.1 M, pH = 6), and then 0.1 mL of linoleic acid sodium salt (0.005 M) was added. The activity of LOX was observed according to the resultant rise in absorbance at 234 nm due to hydroperoxide formation. The activity of LOX was expressed as U min^{-1} mg $protein^{-1}$.

Two equidistant readings were taken in the equatorial region of each fruit. The results were expressed in Newtons (N) [31], and total soluble solid content (SSC%) was measured utilizing Portable Digital Refractometer (RFT-PD Series, Jinan City, China) [70].

4.8. Lipid Peroxidation and Cell Membrane Permeability

Malondialdehyde (MDA) was determined according to the method in [71]. One gram of peach pulp tissue was homogenized with 10 mL of 5% (*w*/*v*) meta-phosphoric acid and 0.2 mL of 2% (*w*/*v*) ethanolic butylated hydroxytoluene (BHT). The resultant mixture was centrifuged at 15,000× *g* rpm for 15 min. Then, 1.0 mL of the resultant supernatant was added to 0.1 mL of 2% (*w*/*v*) BHT, 0.5 mL of 1% (*w*/*v*) of thiobarbituric acid (TBA) in 0.05 M NaOH, and 0.5 mL of 20% (*v*/*v*) hydrochloric acid. The tube mixture was incubated at 95 °C for 30 min. After cooling, the resultant chromogen was extracted by adding 0.8 mL of n-butanol, centrifugation was performed to separate the organic phase, and the absorbance of thiobarbituric acid reactive substances (TBARS) was read at 532 nm. The compound of 1,1,3,3-tetra-ethoxy-propane (Sigma) was used to prepare the calibration curves, and finally, micromoles of MDA per gram FW for the examined samples were estimated. The electrolyte leakage percentage rate (EL%) was evaluated by a conductivity ion meter. Data were exhibited as a percentage [67].

4.9. Ethylene and Respiration

Ethylene and respiration of peach were estimated on a fruit sample of 6 peaches after a 2-day interlude. Fruits were incubated and saved in 1000 mL glass jars sealed for a one-hour interval of the shelf-life term in time for all further analyses. Gas samples only from each of the jars around the fruit were extracted, and both ethylene and carbon dioxide were examined by gas chromatography techniques (GC) (Darmstadt, Germany). However, ethylene was monitored by GC-6000 Vega Series (Carlo Erba Ins., Milano, Italy) and carbon dioxide was measured by GC PBI-Dansensor Checkmate-9900 (Copenhagen, Denmark).

4.10. Statistical Analysis and Data Modeling

The parameters of water loss %, fruit skin color hue angle, and skin spot browning index were analyzed using one-way analysis using a complete block design when GA/PVP–SA treatments were considered as a factor. However, the distractive parameters were analyzed using two-way analysis using a factorial experiment in a completely randomized block design for chemical parameters when storage period (days) and GA/PVP–SA treatments were investigated as factors. The means of all examined treatments were compared using Duncan's multiple range test at $p < 0.05$ using Co-Stat software package Ver. 6.303 (798 lighthouse Ave PMB320, Monterey, CA, USA). To forecast the characteristics of the treated fruits during the shelf life, a linear regression technique was performed, and the criterion of $R^2 > 0.9$ was considered as a proper judge for the fitness of the model [72].

5. Conclusions

The outcomes obtained from this research show that the peach fruit coated with GA/PVP mixed with SA at a concentration of 2 mM gives a significant effect on diminishing fruit tissue breakdown during postharvest life. This result can be summarized into three points: First, the treatment inhibits the activities of CWDEs. Second, it decreases the lipid peroxidation process and ionic penetrability rate. Third, the coating treatment decreases the activities of browning enzymes, and therefore, the brown incidence color on the fruits decreases during the shelf-life period. It can be said that GA/PVP–SA 2 mM treatment can retard the deterioration of peach fruit during shelf life. Future studies for the development of information on the impact of GA/PVP–SA treatment to increase fruit storability under shelf-life conditions are needed.

Author Contributions: Conceptualization, A.A.L., M.A.T. and H.A.K.; methodology, A.A.L., M.A.T., S.A.L., M.A.A., M.F.M.A. and D.M.H.; software, A.A.L., M.A.T., M.F., S.Y.M. and D.M.H.; validation, A.A.L., M.G., M.A.T., M.F., S.Y.M., D.O.E.-A., S.F.E.-G., M.A.A. and D.M.H.; formal analysis, A.A.L., M.A.T., M.G., S.F.E.-G., M.F.M.A. and H.A.K.; investigation, A.A.L., M.A.T., E.F.A., A.M.I., M.F.M., M.A.A. and D.M.H.; data curation, E.F.A., A.A.L., M.G., M.A.T., S.A.L. and M.A.A.; visualization, H.S.G., M.A.A., M.G., S.O.O., A.M.I., M.F.M., D.O.E.-A. and M.F.; writing—original draft preparation, A.A.L., H.A.K., H.S.G., S.O.O., M.F., M.G., M.F.M.A. and M.A.T.; writing—review, and editing, A.A.L., D.O.E.-A., M.G., M.F.M.A., S.A.L., E.F.A. and M.A.T.; funding acquisition, M.F.M.A., A.M.I., M.F.M, S.F.E.-G. and E.F.A. All authors have read and agreed to the published version of the manuscript.

Funding: The authors extend their appreciation to Taif University for supporting this work: Researchers Supporting Project under project number (TURSP-2020/65), Taif University, Taif, Saudi Arabia.

Institutional Review Board Statement: Not applicable.

Informed Consent Statement: Not applicable.

Acknowledgments: The authors are thankful to Taif University for supporting this work: Researchers Supporting Project under project number (TURSP-2020/65), Taif University, Taif, Saudi Arabia.

Conflicts of Interest: The authors declare no conflict of interest.

Sample Availability: Samples of the compounds are not available from the authors.

References

1. Awad, R.M. Effect of post-harvest salicylic acid treatments on fruit quality of peach cv. "Flordaprince" during cold storage. *Aust. J. Basic Appl. Sci.* **2013**, *7*, 920–927.
2. Gil, M.I.; Tomás-Barberán, F.A.; Hess-Pierce, B.; Kader, A.A. Antioxidant Capacities, Phenolic Compounds, Carotenoids, and Vitamin C Contents of Nectarine, Peach, and Plum Cultivars from California. *J. Agric. Food Chem.* **2002**, *50*, 4976–4982. [CrossRef]
3. Gayed, A.A.; Shaarawi, S.A.; Elkhishen, M.A.; Elsherbini, N.M. Pre-harvest application of calcium chloride and chitosan on fruit quality and storability of Early Swelling peach during cold storage. *Ciência E Agrotecnologia* **2017**, *41*, 220–231. [CrossRef]
4. Manganaris, G.A.; Vasilakakis, M.; Diamantidis, G.; Mignani, I. Diverse metabolism of cell wall components of melting and non-melting peach genotypes during ripening after harvest or cold storage. *J. Sci. Food Agric.* **2006**, *86*, 243–250. [CrossRef]
5. Hong, K.; Xie, J.; Zhang, L.; Sun, D.; Gong, D. Effects of chitosan coating on postharvest life and quality of guava (*Psidium guajava* L.) fruit during cold storage. *Sci. Horti.* **2012**, *144*, 172–178. [CrossRef]
6. Guillén, F.; Díaz-Mula, H.M.; Zapata, P.J.; Valero, D.; Serrano, M.; Castillo, S.; Martínez-Romero, D. *Aloe arborescens* and *Aloe vera* gels as coatings in delaying postharvest ripening in peach and plum fruit. *Postharvest Biol. Technol.* **2013**, *83*, 54–57. [CrossRef]
7. Hazrati, S.; Beyraghdar, K.A.; Habibzadeh, F.; Tahmasebi-Sarvestani, Z.; Sadeghi, A.R. Evaluation of *Aloe vera* gel as an alternative edible coating for peach fruits during cold Ssorage period. *Gesunde Pflanz.* **2017**, *69*, 131–137. [CrossRef]
8. Forato, L.A.; de Britto, D.; de Rizzo, J.S.; Gastaldi, T.A.; Assis, O.B.G. Effect of cashew gum-carboxymethylcellulose edible coatings in extending the shelf-life of fresh and cut guavas. *Food Packag. Shelf Life* **2015**, *5*, 68–74. [CrossRef]
9. Lo'ay, A.A.; Dawood, H.D. Minimize browning incidence of banana by postharvest active chitosan/PVA Combines with oxalic acid treatment to during shelf-life. *Sci. Hortic.* **2017**, *226*, 208–215. [CrossRef]
10. Lo'ay, A.A.; Dawood, H.D. Active chitosan/PVA with ascorbic acid and berry quality of 'Superior seedless' grapes. *Sci. Hortic.* **2017**, *224*, 286–292. [CrossRef]
11. Pobiega, K.; Igielska, M.; Włodarczyk, P.; Gniewosz, M. The use of pullulan coatings with propolis extract to extend the shelf life of blueberry (Vaccinium corymbosum) fruit. *Int. J. Food Sci. Technol.* **2021**, *56*, 1013–1020. [CrossRef]

12. Melo, N.F.C.B.; de Lima, M.A.B.; Stamford, T.L.M.; Galembeck, A.; Flores, M.A.; de Campos Takaki, G.M.; da Costa Medeiros, J.A.; Stamford-Arnaud, T.M.; Montenegro Stamford, T.C. In vivo and in vitro antifungal effect of fungal chitosan nanocomposite edible coating against strawberry phytopathogenic fungi. *Int. J. Food Sci. Technol.* **2020**, *55*, 3381–3391. [CrossRef]
13. Zhang, W.; Jing, L.; Chen, H.; Zhang, S. NC-1 coating combined with 1-MCP treatment maintains better fruit qualities in honey peach during low-temperature storage. *Int. J. Food Sci. Technol.* **2022**, *57*, 516–524. [CrossRef]
14. Maqbool, M.; Ali, A.; Alderson, P. A combination of gum arabic and chitosan can control anthracnose caused by *Colletotrichum musae* and enhance the shelf-life of banana fruit. *J. Hortic. Sci. Biotechnol.* **2010**, *85*, 432–436. [CrossRef]
15. Maqbool, M.; Ali, A.; Alderson, P.G.; Zahid, N.; Siddiqui, Y. Effect of a novel edible composite coating based on gum arabic and chitosan on biochemical and physiological responses of banana fruits during cold storage. *J. Agric. Food Chem.* **2011**, *59*, 5474–5482. [CrossRef]
16. Sicari, V.; Loizzo, M.R.; Pellicanò, T.M.; Giuffrè, A.M.; Poiana, M. Evaluation of Aloe arborescens gel as new coating to maintain the organoleptic and functional properties of strawberry (*Fragaria* × *ananassa* cv. Cadonga) fruits. *Int. J. Food Sci. Technol.* **2020**, *55*, 861–870. [CrossRef]
17. Klein, M.; Aserin, A.; Ishai, P.B.; Garti, N. Interactions between whey protein isolate and gum Arabic. *Colloids Surf. B Biointerfaces* **2010**, *79*, 377–383. [CrossRef]
18. Huang, X.; Kakuda, Y.; Cui, W. Hydrocolloids in emulsions: Particle size distribution and interfacial activity. *Food Hydrocoll.* **2001**, *15*, 533–542. [CrossRef]
19. Dickinson, E. Hydrocolloids at interfaces and the influence on the properties of dispersed systems. *Food Hydrocoll.* **2001**, *17*, 25–39. [CrossRef]
20. Renard, D.; Lavenant-Gourgeon, L.; Ralet, M.-C.; Sanchez, C. *Acaciasenegal Gum*: Continuum of molecular species differing by their protein to sugar ratio, molecular weight, and charges. *Biomacromolecules* **2006**, *7*, 2637–2649. [CrossRef]
21. Ali, A.; Maqbool, M.; Ramachandran, S.; Alderson, P.G. Gum arabic as a novel edible coating for enhancing shelf-life and improving postharvest quality of tomato (*Solanum lycopersicum* L.) fruit. *Postharvest Biolo. Technol.* **2010**, *58*, 42–47. [CrossRef]
22. El-Anany, A.M.; Hassan, G.F.A.; Ali, F.M. Effects of Edible Coatings on the Shelf-Life and Quality of Anna Apple (*Malus domestica Borkh*) During Cold Storage. *J. Food Technol.* **2009**, *7*, 5–11.
23. Idris, Y.M.A.; Ibrahim, Y.A.; Mariod, A.A. Color of dehydrated tomato: Effects of gum arabic. *Int. J. Food Prop.* **2013**, *16*, 838–851. [CrossRef]
24. Koczkur, K.M.; Mourdikoudis, S.; Polavarapu, L.; Skrabalak, S.E. Polyvinylpyrrolidone (PVP) in nanoparticle synthesis. *Dalton Trans.* **2015**, *44*, 17883–17905. [CrossRef]
25. Kadlubowski, S. Radiation-induced synthesis of nanogels based on poly(N-vinyl-2-pyrrolidone)—A review. *Radiat. Phys. Chem.* **2014**, *102*, 29–39. [CrossRef]
26. Halake, K.; Birajdar, M.; Kim, B.S.; Bae, H.; Lee, C.; Kim, Y.J.; Kim, S.; Kim, H.J.; Ahn, S.; An, S.Y.; et al. Recent application developments of water-soluble synthetic polymers. *J. Ind. Eng. Chem.* **2014**, *20*, 3913–3918. [CrossRef]
27. Tadda, M.A.; Gouda, M.; Lin, X.; Shitu, A.; Abdullahi, H.S.; Zhu, S.; Li, X.; Liu, D. Impacts of Baobab (Adansonia Digitata) Powder on the Poly(Butylene Succinate) Polymer Degradability to Form an Eco-Friendly Filler-Based Composite. *Front. Mater.* **2021**, *8*, 505. [CrossRef]
28. FDA. Center for Drug Evaluation and Research, Database Update Frequency: Quarterly. CFR—Code of Federal Regulations Title 21: January 06, 2020— Sec. 173.55 Polyvinylpyrrolidone. 2020. Available online: https://www.accessdata.fda.gov/scripts/cdrh/cfdocs/cfcfr/CFRSearch.cfm?fr=173.55 (accessed on 26 March 2022).
29. WHO. *The Ottawa Charter for Health Promotion*; Nov 21; WHO: Geneva, Switzerland, 1986; Available online: http://www.who.int/healthpromotion/conferences/previous/ottawa/en/index.html (accessed on 26 March 2022).
30. Panda, H. *The Complete Book on Gums and Stabilizers for Food Industry*; Asia Pacific Business Press Inc.: Delhi, India, 2010; pp. 1–480.
31. Lo'ay, A.A.; Taher, M.A. Effectiveness salicylic acid blending in chitosan/PVP biopolymer coating on antioxidant enzyme activities under low storage temperature stress of 'Banati' guava fruit. *Sci. Hortic.* **2018**, *238*, 343–349. [CrossRef]
32. Lo'ay, A.A.; Taher, M.A. Influence of edible coatings chitosan/PVP blending with salicylic acid on biochemical fruit skin browning incidence and shelf life of guava fruits cv. 'Banati'. *Sci. Hortic.* **2018**, *235*, 424–436. [CrossRef]
33. El-Banna, M.F.; Lo'ay, A.A. Evaluation berries shattering phenomena of 'Flame seedless' vines grafted on different rootstocks during shelf life. *Sci. Hortic.* **2019**, *246*, 51–56. [CrossRef]
34. Hoff, J.F.; Singleton, K.I. A method for determination of tannin in foods by means of immobilized enzymes. *J. Food Sci.* **1977**, *42*, 1566–1569. [CrossRef]
35. Zhang, X.; Tian, S. Effect of oxalic acid on control of postharvest browning of litchi fruit. *Food Chem.* **2006**, *96*, 519–523. [CrossRef]
36. Lo'ay, A.A.; Rabie, M.M.; Alhaithloul, H.A.S.; Alghanem, S.M.S.; Ibrahim, A.M.; Abdein, M.A.; Abdelgawad, Z.A. On the biochemical and physiological responses of 'Crimson seedless' grapes coated with an edible composite of pectin, polyphenylene alcohol, and salicylic acid. *Horticulturae* **2021**, *7*, 498. [CrossRef]
37. Jiang, Y.; Louce, D.C.; Jayas, W.; Lu, W. Effect of chilling temperatures on ethylene binding by banana fruit. *Plant Growth Regul.* **2004**, *43*, 109–115. [CrossRef]
38. Chen, Y.; Sun, J.; Lin, H.; Lin, M.; Lin, Y.; Wang, H.; Hung, Y.-C. Salicylic acid reduces the incidence of *Phomopsis longanae* Chi infection in harvested longan fruit by affecting the energy status and respiratory metabolism. *Postharvest Biol. Technol.* **2020**, *160*, 111035. [CrossRef]

39. Pathak, N.; Sanwal, G.G. Multiple forms of polygalacturonase from banana fruits. *Phytochemistry* **1998**, *48*, 249–255. [CrossRef]
40. Belyakova, L.A.; Varvarin, A.M.; Lyashenko, D.Y.; Khora, O.V.; Oranskaya, E.I. Complexation in a b-Cyclodextrin–Salicylic Acid System. *Colloid J.* **2007**, *69*, 546–551. [CrossRef]
41. Bashir, M.; Haripriya, S. Assessment of physical and structural characteristics of almond gum. *Int. J. Biol. Macromol.* **2016**, *93*, 476–482. [CrossRef]
42. Mireles, L.K.; Wu, M.-R.; Saadeh, N.; Yahia, L.H.; Sacher, E. Physicochemical Characterization of Polyvinyl Pyrrolidone: A Tale of Two Polyvinyl Pyrrolidones. *ACS Omega* **2020**, *5*, 30461–30467. [CrossRef]
43. Pérez, A.G.; Sanz, C.; Olías, R.; Olías, J.M. Lipoxygenase and Hydroperoxide Lyase Activities in Ripening Strawberry Fruits. *J. Agric. Food Chem.* **1999**, *47*, 249–253. [CrossRef]
44. Srivastava, M.K.; Dwivedi, U.N. Delayed ripening of banana fruit by salicylic acid. *Plant Sci.* **2000**, *158*, 87–96. [CrossRef]
45. Meindrawan, B.; Ofe, O.; Susanto, C.S.; Ayman, A.; Mangindaan, D.; Kasih, T.P. Glucomannan–beeswax–chitosan antimicrobial edible coating to maintain the storage quality of salak fruit. *Salacca Zalacca* **2020**, *391*, 1900164. [CrossRef]
46. Lo'ay, A.A.; Doaa, M.H. The potential of vine rootstocks impacts on 'Flame Seedless' bunches behavior under cold storage and antioxidant enzyme activity performance. *Sci. Hortic.* **2020**, *260*, 108844. [CrossRef]
47. Meng, K.; Hou, Y.; Han, Y.; Ban, Q.; He, Y.; Suo, J.; Rao, J. Exploring the Functions of 9-Lipoxygenase (DkLOX3) in Ultrastructural Changes and Hormonal Stress Response during Persimmon Fruit Storage. *Int. J. Mol. Sci.* **2017**, *18*, 589. [CrossRef] [PubMed]
48. Tadda, M.; Gouda, M.; Lin, X.; Shitu, A.; Abdullahi, H.; Zhu, S.; Li, X.; Liu, D. Evaluation of *Aloe vera* gel as an alternative edible coating for peach fruits during cold storage period. *Front. Mater.* **2021**, *8*, 768960. [CrossRef]
49. Gouda, M.; El-Din Bekhit, A.; Tang, Y.; Huang, Y.; Huang, L.; He, Y.; Li, X. Recent innovations of ultrasound green technology in herbal phytochemistry: A review. *Ultrason. Sonochem.* **2021**, *73*, 105538. [CrossRef]
50. McCann, M.C.; Carpita, N.C. Designing the deconstruction of plant cell walls. *Curr. Opin. Plant Biol.* **2008**, *11*, 314–320. [CrossRef]
51. Atkinson, R.G.; Schröder, R.; Hallett, I.C.; Cohen, D.; MacRae, E.A. Overexpression of Polygalacturonase in Transgenic Apple Trees Leads to a Range of Novel Phenotypes Involving Changes in Cell Adhesion. *Plant Physiol.* **2002**, *129*, 122–133. [CrossRef]
52. Qian, M.; Zhang, Y.; Yan, X.; Han, M.; Li, J.; Li, F.; Li, F.; Zhang, D.; Zhao, C. Identification and Expression Analysis of Polygalacturonase Family Members during Peach Fruit Softening. *Int. J. Mol. Sci.* **2016**, *17*, 1933. [CrossRef]
53. Goyal, A.; Ghosh, B.; Eveleigh, D. Characteristics of fungal cellulases. *Bioresour. Technol.* **1991**, *36*, 37–50. [CrossRef]
54. Abu-Goukh, A.-B.A.; Bashir, H.A. Changes in pectic enzymes and cellulase activity during guava fruit ripening. *Food Chem.* **2003**, *83*, 213–218. [CrossRef]
55. Wang, D.; Zhang, H.; Wu, F.; Li, T.; Liang, Y.; Duan, X. Modification of Pectin and Hemicellulose Polysaccharides in Relation to Aril Breakdown of Harvested Longan Fruit. *Int. J. Mol. Sci.* **2013**, *14*, 23356–23368. [CrossRef] [PubMed]
56. Dave, R.K.; Ramana Rao, T.V.; Nandane, A.S. Improvement of post-harvest quality of pear fruit with optimized composite edible coating formulations. *J. Food Sci. Technol.* **2017**, *54*, 3917–3927. [CrossRef] [PubMed]
57. Lo'ay, A.A.; Elgammal, R.E.; Alhaithloul, H.A.S.; Alghanem, S.M.; Fikry, M.; Abdein, M.A.; Hikal, D.M. Enhance Fruit Ripening Uniformity and Accelerate the Rutab Stage by Using ATP in 'Zaghloul' Dates during the Shelf Life. *Foods* **2021**, *10*, 2641. [CrossRef]
58. Zhu, S.; Zhou, J. Effects of Nitric Oxide on Fatty Acid Composition in Peach Fruits during Storage. *J. Agric. Food Chem.* **2006**, *54*, 9447–9452. [CrossRef]
59. Wang, C.; Fu, M.; D'Amico, M.; Albanese, C.; Zhou, J.-N.; Brownlee, M.; Lisanti Michael, P.; Chatterjee, V.K.K.; Lazar Mitchell, A.; Pestell Richard, G. Inhibition of Cellular Proliferation through IκB Kinase-Independent and Peroxisome Proliferator-Activated Receptor γ-Dependent Repression of Cyclin D1. *Mol. Cell. Biol.* **2001**, *21*, 3057–3070. [CrossRef]
60. Hildebrand, D.F. Lipoxygenases. *Physiol. Plant* **1989**, *76*, 249–253. [CrossRef]
61. Murmu, S.B.; Mishra, H.N. The effect of edible coating based on Arabic gum, sodium caseinate and essential oil of cinnamon and lemon grass on guava. *Food Chem.* **2018**, *245*, 820–828. [CrossRef]
62. Petriccione, M.; Mastrobuoni, F.; Pasquariello, M.S.; Zampella, L.; Nobis, E.; Capriolo, G.; Scortichini, M. Effect of Chitosan Coating on the Postharvest Quality and Antioxidant Enzyme System Response of Strawberry Fruit during Cold Storage. *Foods* **2015**, *4*, 501–523. [CrossRef]
63. Babalar, M.; Asghari, M.; Talaei, A.; Khosroshahi, A. Effect of pre- and postharvest salicylic acid treatment on ethylene production, fungal decay and overall quality of Selva strawberry fruit. *Food Chem.* **2007**, *105*, 449–453. [CrossRef]
64. Li, N.; Parsons, B.L.; Liu, D.R.; Mattoo, A.K. Accumulation of wound-inducible ACC synthase transcript in tomato fruit is inhibited by salicylic acid and polyamines. *Plant Mol. Biol.* **1992**, *18*, 477–487. [CrossRef] [PubMed]
65. Fan, X.; Mattheis, J.P.; Fellman, J.K. Inhibition of apple fruit 1-aminocyclopropane-1-carboxylic acid oxidase activity and respiration by acetylsalicylic acid. *J. Plant Physiol.* **1996**, *149*, 469–471. [CrossRef]
66. Lo'ay, A.A.; Ismail, H.; Kassem, H.S. Postharvest Treatment of 'Florida Prince' Peaches with a Calcium Nanoparticle–Ascorbic Acid Mixture during Cold Storage and Its Effect on Antioxidant Enzyme Activities. *Horticulturae* **2021**, *7*, 499. [CrossRef]
67. Lo'ay, A.A. Chilling Injury in Mangoes. Ph.D. Thesis, Wageningen University, Wageningen, The Netherlands, 2005; pp. 1–224.
68. Lohani, S.; Triverdi, P.K.; Nath, P. Changes in activities of cell wall hydrolases during ethylene-induced ripening in banana: Effect of 1-MCP, ABA and IAA. *Postharvest Biol. Technol.* **2004**, *31*, 119–126. [CrossRef]
69. Bradford, M.M. A rapid and sensitive method for the quantitation of microgram quantities of protein utilizing the principle of protein-dye binding. *Anal. Biochem.* **1976**, *72*, 248–254. [CrossRef]

70. Association of Official Agricultural Chemists; Horwitz, W. *Official Methods of Analysis*; Association of Official Analytical Chemists: Washington, DC, USA, 2005.
71. Iturbe-Ormaetxe, I.; Escuredo, P.R.; Arrese-Igor, C.; Becana, M. Oxidative damage in pea plants exposed to water deficit or paraquat. *Plant Physiol.* **1998**, *116*, 173–181. [CrossRef]
72. Fikry, M.; Aniza Yusof, Y.; Al-Awaadh, A.M.; Abdul Rahman, R.; Ling Chin, N.; Mousa, E.; Sin Chang, L. Kinetics Modelling of the Colour, Hardness, Grinding Energy Consumption and Oil Yield Changes during the Conventional Roasting of Palm Date Seeds. *Food Sci. Technol. Res.* **2019**, *25*, 351–362. [CrossRef]

molecules

Review

The Recent Development of Acoustic Sensors as Effective Chemical Detecting Tools for Biological Cells and Their Bioactivities

Mostafa Gouda [1,2,*], Hesham S. Ghazzawy [3,4,*], Nashi Alqahtani [3] and Xiaoli Li [1,*]

1 College of Biosystems Engineering and Food Science, Zhejiang University, 866 Yuhangtang Road, Hangzhou 310058, China
2 Department of Nutrition & Food Science, National Research Centre, Dokki, Giza 12622, Egypt
3 Date Palm Research Center of Excellence, King Faisal University, Al Ahsa 31982, Saudi Arabia
4 Central Laboratory for Date Palm Research and Development, Agriculture Research Center, Giza 12511, Egypt
* Correspondence: mostafa-gouda@zju.edu.cn or goudarowing@yahoo.com (M.G.); hghazzawy@kfu.edu.sa (H.S.G.); xiaolili@zju.edu.cn (X.L.)

Abstract: One of the most significant developed technologies is the use of acoustic waves to determine the chemical structures of biological tissues and their bioactivities. In addition, the use of new acoustic techniques for in vivo visualizing and imaging of animal and plant cellular chemical compositions could significantly help pave the way toward advanced analytical technologies. For instance, acoustic wave sensors (AWSs) based on quartz crystal microbalance (QCM) were used to identify the aromas of fermenting tea such as linalool, geraniol, and trans-2-hexenal. Therefore, this review focuses on the use of advanced acoustic technologies for tracking the composition changes in plant and animal tissues. In addition, a few key configurations of the AWS sensors and their different wave pattern applications in biomedical and microfluidic media progress are discussed.

Keywords: acoustic sensors; quartz crystal microbalance; emerging technologies; piezoelectric materials; acoustic wave devices

1. Introduction

The development of acoustic sensors is a very important scientific and technical issue. Acoustic sensors are widely applied in various technical systems that monitor the environment, provide biological and chemical safety, are used in signal processing devices, and numerous other applications. This technology is based on the use of acoustic wavenumbers, which has become one of the important emerging technologies [1,2].

Moreover, acoustic wave devices have been commercially used in many fields for several decays. Several of the emerging applications for acoustic wave devices as sensors include industrial and medical applications (such as tracking the industrial lines' vapor, humidity, temperature, and product quality). For instance, super high sensitivity to humidity has been shown in sensors based on plate acoustic waves with graphene oxide sensitive film [3]. That is due to sensitivity of these sensors and intrinsic reliability. Virtually all acoustic wave devices and sensors are using piezoelectric materials to generate the acoustic wave. These piezoelectricity sensors are made of quartz materials that could have the resonator ability for stabilizing electronic oscillators [4]. Piezoelectricity refers to the production of electrical charges by the imposition of mechanical stress. Additionally, applying an appropriate electrical field to a piezoelectric material creates mechanical stress through an oscillating electric field, which propagates through the substrate and is then converted back to an electric field for measurement [5].

For example, the expression of AWSs occurs due to their detection mechanism based on the mechanical movements that cause acoustic waves to be considered and characterized [6]. These waves are moved on the surface of the material or through them. Any changes in the material characteristics could affect the velocity and/or amplitude of the hit acoustic waves and then could be sensitively detected and correlated to the corresponding physicochemical material measured [7,8].

This review provides the latest achievements related to the design, fabrication, modeling, testing, characterization, and advanced research trends in acoustic sensor technology as efficient analytical methods.

2. Operating Principles of Various Acoustic Sensor Types

The chemical sensors based on acoustic wave technology have continuously received research and technological attention [9,10]. Among the important advantages of the AWS sensors are their ultra-high sensitivity, excellent response time, small size, excellent selectivity, stability, and their compatibility with other emerging sensation technologies such as interdigital transducers (IDTs) [11]. The principle of sound radiation originated from vibrating plane surfaces (VPSs). The responsible detector can discriminate the generated harmonic transverse sound waves that move along an infinite plane surface in contact with a fluid to derive an expression for the associated acoustic radiation impedance [12]. These waves oscillate subsonically with wavenumbers higher than the acoustic wavenumber at the normally considered frequencies, but have only a very low extension into the fluid surface. On the other hand, waves traveling supersonically that have wavenumbers less than the normal acoustic wavenumber generate plane-traveling waves in the fluid that transport energy to infinite distances. This form of analysis can be extended to arbitrary distributions of plane surface vibration utilizing spatial wavenumber spectra for each component that represents the harmonic traveling waves of the studied samples. The sound fields generated by each wavenumber component are then pooled to provide the total radiated field characteristics. Additionally, this form of analysis has computational and interpretational advantages which are used to image the sources based on their reflected and radiated sound fields [13–16]. For example, Surface acoustic wave (SAW) and Bulk acoustic wave (BAW) are mainly used in the field of analytical measurements [17].

2.1. *Surface Acoustic Wave (SAW) Sensors: Different Types of Devices*

Surface acoustic wave (SAW) is a specific kind of acoustic wave that travels down a material's surface at a depth of roughly one or two wavelengths. Since the majority of the SAW energy is concentrated in the vicinity of the surface, the piezoelectric material's surface is highly sensitive to even the smallest perturbations [18]. The intensities, phase angles, and output frequencies of the acoustic waves may change due to variations in wave velocity and attenuation brought on by acoustoelectric interactions or the mass loading effect, which can be observed using detection equipment [9]. Thus, these sensors are expected to fulfill the increasing demand for fast and sensitive detecting and monitoring technologies for various organic and inorganic materials (Figure 1). In addition, the most used elements in fabricating piezoelectric acoustic sensors are quartz, lithium niobate, gallium arsenide, silicon carbide, zinc oxide, aluminum nitride, graphene oxide, and lithium tantalate [3,19–21] (Table 1).

Table 1. Details of different applications, along with their advantages and disadvantages, in the field of acoustic devices.

No	Type	Samples	Acoustic Frequency	Active Area	Application	Advantages	Disadvantages	Ref
1	Chromium-coated QCMs	Polyclonal goat anti-rabbit IgG	5 MHz	10 mm^2	Measurement of antigens	Simplicity, small volumes of sample, free of expensive reagents, easy fluid control in dynamic mode, and low cost.	QCM sensors have complex circuitry, poor signal-to-noise ratio, and can be influenced by humidity.	[22]
2	$LiTaO_3$-based Love wave sensor	Animal stem cells	128 MHz	10 × 12 mm	Quantitative measurements of cell activities	Real-time measurement and propagation could be controlled through the used quartz crystal, allowing a simple, non-invasive, and quantitative measurement of the adherent cells' viscoelastic properties.	Lack of experiment-related discussion, and there is little research focusing on the theoretical modeling of cell-based Love wave sensors and in-depth comprehensive theoretical analysis.	[23]
3	Piezoelectric lithium niobate ($LiNbO_3$)	Pancreatic cancer diagnosis	28.3 MHz	16 × 40 mm	microRNA and oligonucleotide	Label-free, specific on-chip detection of RNA is achieved by using a separate device.	The lysis rate was only 38%. Significant improvements are needed for optimizing the time.	[24]
4	AT cut quartz substrate	Spirulina (*Arthrospira platensis*) cells	117 MHz	--	Measurement of cadmium and mercury heavy metals	The detection limit was determined to be 10^{-14} M for each metal. Not specific to a single metal, but it provides a global response to the presence of heavy metals in a trace amount.	Should be applied with other micro-organisms for achieving other toxicity tests.	[25]
5	AT-cut quartz substrate	*Escherichia coli* (*E. coli*) bacteria	118 MHz	10 mm^2	Measurement of bacteria antigen/antibody reactions	Offer a great and specific affinity especially when a monoclonal antibody is used through detecting the specific interaction between that antibody and the antigen.	The limitation of the solid/liquid interface of the sensor with the higher biological environment to the limited sensing area of 10 mm^2.	[26,27]
6	Quartz crystal	Bovine and pig tissues	392 MHz	25 × 2.5 mm^2	Odorant-binding proteins	The high sensitivity of 5.63 Hz/ppm was obtained with a detection limit of 1.78 ppm with high reproducibility.	Needs low viscosities liquids for uniformly coating the active area of SAW resonators.	[28,29]
7	YX-LiTaO3 crystals	Plasma/serum	426.4 MHz	4 × 4 mm	C-reactive protein (CRP)	It successfully distinguished the human CRP serum normal concentrations from the bacterially infected tissue injury.	The limited binding ability of CRP is based on the period. In addition, especially in serum samples, the adsorption capacity was different among the samples and, consequently, signal responses were affected.	[13,30]
8	Y-cut X-rotated lithium niobate (LN)	Protein solutions	900 MHz	5 mm diameter holes	Biotin-polyethylene glycol-thiol and streptavidin	The sensitivity was 296 $m^2\ kg^{-1}$ and the limit of detection was 104 pM. Adapted for cancer biomarker detection.	Sensitivity is still lower than those based on optical detection.	[31]

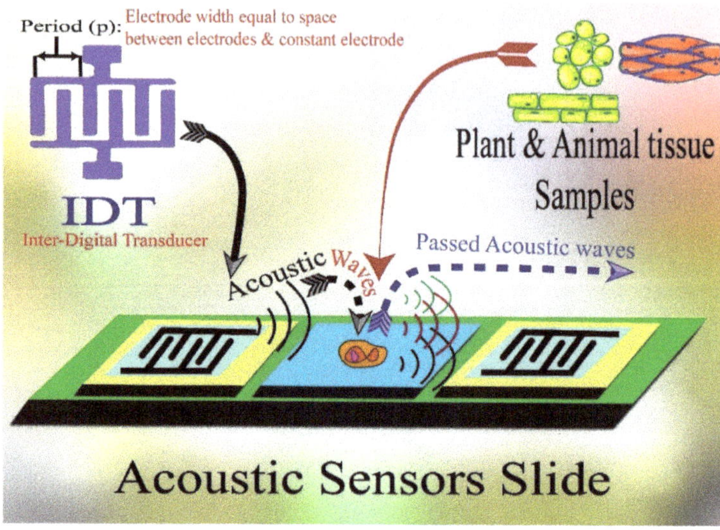

Figure 1. Schematics of SAW chemical sensors: a two-port delay line and a resonator with sensing overlayers for the target analyte.

In addition, there are several SAW propagation types, such as Rayleigh waves, Sezawa waves, surface transverse waves, and shear horizontal modes. The sensor should have elements with high biological affinity such as gold (Au) to be adopted for the cell's extra adhesion layer [32]. Each has specific advantages and disadvantages, which include cost, temperature dependence, attenuation, and propagation velocity. An interesting property of quartz is that it is possible to select the temperature dependence of the material by the cut angle and the wave propagation direction. Other materials with commercial potential include gallium arsenide, silicon carbide, zinc oxide, and aluminum nitride.

2.1.1. Love Acoustic Wave Sensor Definition and Working Principle

Love waves are surface waves with a horizontal motion that propagate shear mode waves supported on semi-infinite substrates with a waveguide layer that exhibits a shear acoustic speed lower than that of the substrate (Figure 2). It is well known that a Love wave has a pure shear horizontal polarization and a small attenuation of the wave is caused in liquid media. On the other hand, the other acoustic wave sensors utilize IDT to generate and detect propagating shear modes that polarized the surface acoustic waves and surface transverse waves. Love-mode acoustic devices are very promising as biosensors in gaseous and liquid environments because of their high sensitivity. For example, Du et al. [33] developed a Love-wave device based on SiO_2/ST-cut quartz over a wide range of SiO_2 thicknesses. In that study, the authors used devices with up to 7.3 μm thick SiO_2, and they measured the mass sensitivity, velocity, insertion loss, oscillation frequency stability, and temperature coefficient of the frequency of the different functional layers. In addition, the authors reported that high sensitivity ($\geq$300 cm^2 g^{-1}) can be achieved at a thickness between 3.5 and 6.5 μm of the quartz layer. Thus, numerical modeling of device thickness, mass, and liquid response should be specifically directed to a particular Love wave device type.

The Love wave sensor directly measures cell/substrate bonds via acoustic damping and provides 2D kinetic and affinity parameters. Other studies have applied the QCM sensor as a diagnostic tool for leukemia and, potentially, for chemotherapeutic agents. Acoustic sensors have also been used in the evaluation of the cytocompatibility of artificial surfaces and, in general, they have the potential to become powerful tools for even more diverse cellular analysis [25,34].

In another study of Love wave acoustic biosensors for monitoring the adhesion process of stem cells [23], the authors reported that Love wave biosensor is considered to be one of the most promising probing methods in biomedical research and diagnosis fields. It can detect the mechano-biological behavior of cells attached to the surface of the device. In that study, a lithium tantalate ($LiTaO_3$)-based Love wave sensor was adopted as a cell-based biosensor to monitor the adhesion process. The effects of the viscoelastic cell layer and waveguiding layer on the propagation velocity v and propagation loss (PL) were investigated. The resulting different storage and loss shear modulus revealed the potential of that technology's usefulness in quantitative measures of cellular activities under multiple physiological conditions.

2.1.2. Shear Horizontal Acoustic Wave Definition and Working Principle

Shear horizontal acoustic sensors are propagating modes involving the thickness of a thin piezoelectric plate and the detection and excitation by IDT. In 1968, Bleustein discovered shear horizontal surface acoustic waves (SH-SAW) based on a barium titanate ($BaTiO_3$) piezoelectric material [35]. SH-SAW propagates due to the vertical movement of surface particles relative to the wave propagation direction and surface normality, as presented in Figure 2a. In terms of sensing capabilities, SH-SAW sensors are generally used for the detection of liquid samples due to less energy dissipating into the liquid during wave propagation, Figure 2d. In addition, there are many available materials for SH-SAW sensor substrates such as quartz, lithium tantalate ($LiTaO_3$), and lithium niobate ($LiNbO_3$), as well as other crystals such as langasite. The SH-modes in SH-APM resonators can be considered as superposition of plane waves with in-plane displacement reflected at a particular angle between the upper and lower face of the quartz resonator involving the full thickness of the resonator. This kind of sensor has very important applications in the analytical microbial pathogenicity and the responsible chemical measurements. For instance, Ji et al. [36] used the aptamer-based shear horizontal surface acoustic wave biosensor single-layered graphene film for high-sensitivity detection of the *Escherichia coli* cells endotoxin. These endotoxins are complex lipopolysaccharides that are produced from the cell walls of various Gram-negative bacteria (possessed lipid, core polysaccharide, and O-polysaccharide side chains), which are important molecules in measuring bacterial toxicity.

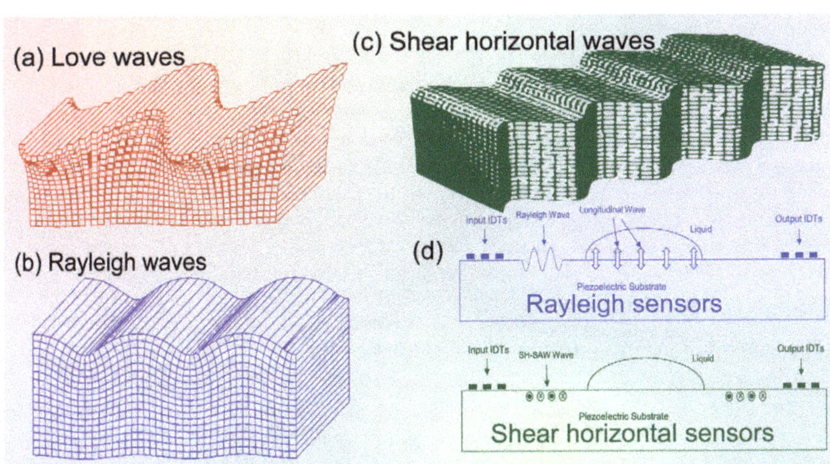

Figure 2. Schematics of different surface acoustic waves. (**a**) Love waves, (**b**) Rayleigh waves, (**c**) Shear horizontal (SH) waves [17,37]. (**d**) Schematic diagram showing Rayleigh wave and SH-SAW sensors' application principles. The particle movement in and out is indicated by "cross: ×" and "bold dot: •" labels.

2.1.3. Rayleigh Acoustic Wave Definition and Working Principle

Rayleigh waves are considered surface waves that travel only along a free surface or along the boundary between two dissimilar solid and liquid media (Figure 2). The name "Rayleigh" came from the British physicist Lord Rayleigh who discovered SAWs in 1885 [38]. Rayleigh waves are surface-normal waves, which makes them unsuitable for use in deep liquid sensing devices. Rayleigh waves are formed when the particle motion is a combination of both longitudinal and transverse vibration giving rise to an elliptical retrograde motion in the vertical plane along the direction of travel. This technology has a vast application for the biochemical analyses of biological fluids. For example, Agostini et al. [31] used a label-free sub-nanomolar Rayleigh surface acoustic-wave (R-SAW)-based biosensor in demonstrating the biomolecular detection of dried liquid solids. That biosensor integrated two interdigital transducers for positive and negative reflectors. The experiments demonstrate a limit of detection of 104 pM and a normalized sensitivity of -296 m^2 kg^{-1}. In comparison with similar acoustic-wave-based systems, both the sensitivity and limit of detection of that sensor were higher than those of Love-SAW biosensors in its application in cancer biomarker detection. Additionally, the velocity of propagation of any body wave in any homogeneous, isotropic material is determined by the elastic moduli and densities of the material through which it passes. The traditional seismic survey uses only compressional waves due to easy detection of the vertical ground motion in the detectors that becomes fast because of high-speed wave velocity. On the other hand, the recording of stress and Rayleigh surface waves provides greater information about the subsurface, but at a cost of greater data acquisition and consequent complex processing [22].

2.1.4. Quartz Crystal Microbalance (QCM) Definition and Principle

Quartz crystal microbalances are suitable transducers for chemical and biochemical sensing [28]. They are used to detect the micro mass changes and physical properties of thin layers deposited on the crystal surfaces and are capable of real-time detection [39]. These devices are using silicon dioxide (SiO_2) known as quartz and have higher shear wave velocity and higher density than polymer materials such as polymethylmethacrylate (PMMA), providing the potential for lower acoustic loss [40]. Due to its simplicity, small volumes of samples, absence of expensive reagents, easy fluid control in the dynamic mode, and low cost, it has been reported as important for detection of dynamic biomolecules.

Several studies have reported on QCM biosensors as being an alternative method to conventional analytical methods such as chromatography and immunosorbent assays [22]. One of the important advantages of using these sensors is that they are reusable, with high sensitivity. Matatagui et al. [22] developed an immunosensor based on a QCM device with very high sensitivity for the detection of antigens in real-time status. In that study, the QCM immunosensor was well-established for the detection of immunological reactions. The biosensor comprises a quartz crystal with an antigen or antibody immobilized on its surface. The reaction between antigen and antibody is thus promoted due to the antigens being transported by the movement of the liquid medium. Moreover, immunoreactions occur due to diffusion of the antigens.

2.1.5. Sezawa and Leaky Pseudo-Acoustic Mode Definition and Principle

From all the studied types of acoustic waves, Sezawa SAW has high acoustic velocity and a large electromechanical coupling coefficient that make it useful for elucidating the bio-sensing properties [41]. It has the highest resonant frequency of up to 17.7 GHz and a signal amplitude of 20 dB with an electromechanical coefficient equal to 0.92% [42]. These devices have shown promising characteristics in many applications due to their compact structures, low power consumption, easy construction and simple packaging, high sensitivity, and fast response that can be used for diverse applications such as sensing ultraviolet radiation, gas, humidity, pH, and biomolecules. For instance, Kuznetsova et al. [43] mentioned that the Sezawa wave can be used to develop highly sensitive humidity acoustic sensors. On the other hand, Suenaga et al. [44] mentioned that leaky pseudo-SAW (LPSAW) is a high-order mode that radiates its acoustic energy into both the water and the substrate. In addition, the water medium is assumed to be an ideal liquid for this mode of measurement.

2.2. Bulk Acoustic Wave Sensor Definitions and Working Principles

The basic design of a BAW resonator consists of a piezoelectric material sandwiched between two metal electrodes. It works on the inverse piezoelectric effect when the electric field is applied across the metal electrode, thereby exciting the acoustic wave in the direction of thickness and vice versa. Zou et al. [45] studied the potential application of thin-film bulk acoustic wave resonators and vigorously supported aluminum nitride (AlN) as a piezoelectric device. In that study, the authors were able to achieve high-quality factor (QF) by reflecting the bulk acoustic wave with air interface at the bottom and top surfaces of resonators. Bulk acoustic waves propagate in a vertical direction with a high-frequency resonance signal, in which different materials and structures of BAW are utilized for optimizing the resonance of acoustic waves between the top and bottom electrodes [46]. Gomes [47] reported that BAW devices have been developed for the quantification detection of a large number of compounds, such as organic compounds, pollution, and biomarkers [48].

3. A Comparative Analysis between BAW and SAW

As SAW was invented almost four decades before BAW devices, both of these sensors have a multitude of measurements in physical, chemical, and biological fields [49]. Both of these sensors are mainly based on oxide ceramics and metals such as quartz. Their output signals such as frequency and phase lend themselves well to digital measurement; they are typically operated at high frequencies [17]. The use of SAW and BAW for chemical composition measurement to draw chemical images of animal and plant cells and tissues and visualize their biomolecules has become one of the hot scientific research areas [50]. An acoustic wave sensor typically consists of a piezoelectric substrate coated with sensing material (polymeric film), and input and output transducers are commonly used for chemical composition purposes. The difference between SAW and BAW is based on the acoustic wave propagation direction: if it moves on the surface of the substrate, it is called SAW, while if the wave propagates through the substrate, it is called BAW (Figure 3).

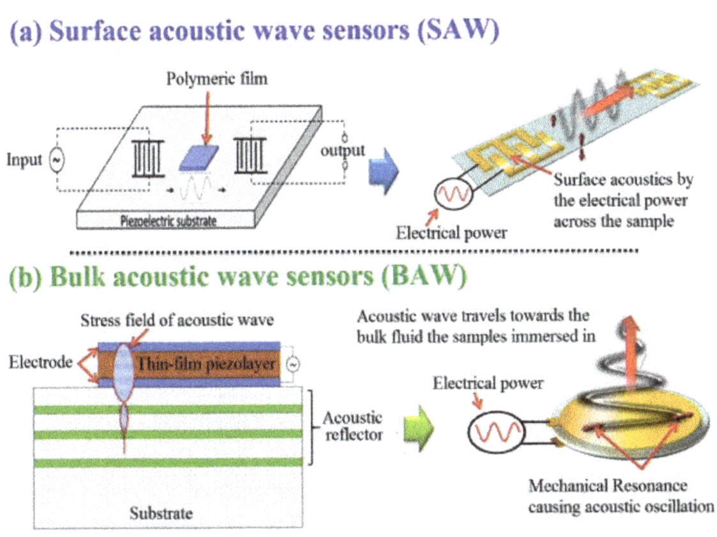

Figure 3. Graphic depicting in general terms the processes for the generation of surface and bulk acoustic waves (open access free permission) [50].

4. Characterization of Electrophysical Properties of Acoustic Sensors

The mode of wave propagation by a piezoelectric substrate is used to characterize acoustic wave devices. The majority of the energy density is contained in the vicinity of the surface due to the order of wavelength penetration depth of these waves [51]. Because of this, any physical or chemical changes on or near the surface causes the waves to alter, and, consequently, the devices that rely on them [52]. For instance, Kiontke et al. [53] used SAW nebulization assistance in substantial signal enhancement. In that study, the authors mentioned that SAW increased the sensitivity response of aminophenols and phenylenediamines up to eight times without any heating for the studied sample due to the increase in the accessible droplet surface area. In addition, the working frequency of SAW devices can be adjusted across a large range (from MHz to GHz), making it possible to fine-tune their sensitivity and use them wirelessly. Additionally, based on the material and boundary conditions, many combinations of velocities and displacement directions can be used to distinguish acoustic waves. As another characteristic example, bulk waves are waves that propagate through the substrate [54]. The thickness shear mode (TSM) resonator and the shear horizontal acoustic plate mode (SH-APM) sensor are the two most widely utilized bulk acoustic wave (BAW) devices (Figure 4).

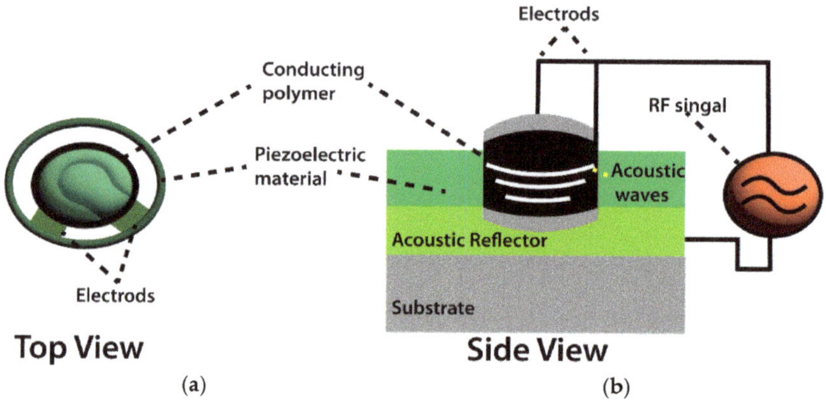

Figure 4. Schematics of bulk acoustic wave (BAW) chemical sensors. (**a**) Vertical top view. (**b**) Side view of BAW sensor system [54,55].

5. Design of Various Types of Acoustic Sensors

The design of the used ASW is based on the used samples' characteristics and nature. For instance, SAW sensors could typically run between 25 and 500 MHz of the acoustic wave frequencies [56]. The surface wave is excessively attenuated when liquid comes in contact with a SAW sensor because of the ensuing compressional waves. The sensitivity of the sensor is often inversely correlated with the energy perturbing the propagation channel [57]. Meanwhile, the energy is often transferred from one surface to the other surface by way of the bulk material in bulk acoustic wave sensors [54]. In addition, the energy density on the surface, which is where the sensing is carried out, is reduced by this energy dispersal. Moreover, other design considerations when selecting acoustic wave sensors include oscillator stability and noise level [57]. These sensors are capable of identifying and detecting substances at ppb levels of concentration [58]. The piezoelectric transducer characteristics, the center frequency, the sensor layer characteristics (such as material qualities, thickness, and surface roughness), and the operating temperature all affect sensitivity, which can be measured in Hz/ppm or Hz/% [59]. The mass sensitivity (f/m) of these characteristics has been demonstrated in numerous tests to rise the root mean square (RMS) of the operating frequency that enhances the signal sensitivity. In ideal circumstances, sensitivity rises with layer thickness; however, variations in a layer's roughness, crystallinity, and hardness with layer thickness could have an impact on sensitivity [32]. Nevertheless, the vast majority of acoustic sensors are Si-based products, which means that they lack several essential elements for the biological sample, including macro- and microelements, such as essential amino acids and fatty acids. Therefore, the acoustic sensor is a great vehicle for fortification with highly valued technological ingredients. For instance, Jiang et al. [60] used SAW sensors with Love acoustic waves by employing SiO_2-coated piezoelectric for detecting hemagglutinin (HA) antibodies that related to Influenza antigen detection with appropriate surface functionalization. The authors mentioned that the HA detection limit concentration is as low as 1 ng mL^{-1}.

The adequate enrichment of acoustic sensors is more effective to improve their detection sensitivity. In this context, chemical analysis technologies should serve with the technological studies to achieve the goal of adding the beneficial essential elements to improve the stability and accuracy of the acoustic sensor for enhancing its feasibility as a worldwide distributed functional chemical analysis technology [61]. For example, the acoustic sensor enriched with powder bacterial cellulose showed benefits in many aspects such as high-sensitivity characteristics [62]. Meanwhile, SAW torque sensors are utilized in practical applications with their centerlines at right angles and their sensitivity to temperature drift; the mass loads make several applications possible for these kinds of sensors, such as their film thickness sensors and particle sensors. The sensor transforms into a particulate sensor if it is covered with an adhesive substance; any particle that lands on the surface stays there and disrupts wave propagation. A 200 MHz ST-cut quartz SAW has been reported to have a mass resolution of 3 pg, which is 1000 times more sensitive than the tested 10 MHz TSM resonator. Wang, Guo, Li, Long, Tang, Zu, Ma, Du, Tang, Torun and Fu [62] developed a SAW for tracking the media humidity as a sensor based on bacterial cellulose (BC) coated by quartz. The authors mentioned that the BC-SAW sensor exhibited good short-term repeatability and long-term stability for the media humidity sensation (Figure 5).

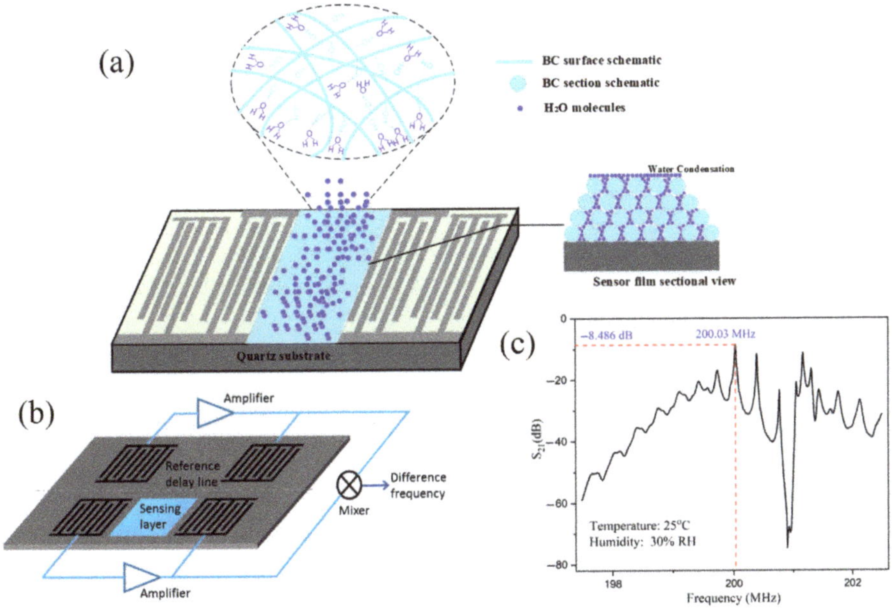

Figure 5. Schematic of a dual delay line SAW sensor for compensation of thermal and humidity drifts. (**a**) SAW by using the Bacterial cellulose (BC) mode of action. (**b**) The sensation layer and reference layer complete SAW system. (**c**) The potential obtained spectrum from using different acoustic frequencies [54,58,62] (copyright permission no: 6895891235874).

6. Application of Acoustic Sensors in Biochemical Material Detection

The application of acoustic wave sensors for monitoring the frequency variations in waves that pass through biochemical materials makes it a perfect technology for identifying the biomarker compounds such as lipids and proteins (Table 1). For instance, it can identify cancer proteins that are bound to a sensor surface receptor [18] (Figure 6a). The piezoelectric effect, a phenomenon where an initial electric signal is transformed into a mechanical displacement, is what causes the initial wave to be produced. This movement is transmitted through the crystal as a wave. In addition, the signal in the SAW sensor moves through the material from the input transducer to the output transducer, where it is transformed back into an electrical signal. According to the Center for Nanoscale Materials (Zhejiang University, China), the wave's frequency is dictated by the sound wave's speed through the material. The capacity of researchers to identify frequency or variations in the waves as they spread is what makes these gadgets effective as sensors [63]. The attachment of chemicals to crystal receptors or proteins to antigens results in changes in the density of the crystalline medium, which in turn induces variations in pitch are an important factor to be considered related to this technology. According to Mandal and Banerjee [64], when something binds to the acoustic sensing layer, the wave characteristics change, then the detector can measure those changes. These novel sensor theories have several advantages, one of which is the possibility of making them battery-operated and portable. However, in order to do that, researchers must figure out how to use less energy to run the apparatus. The ability to run with very low power consumption is necessary to make something portable, according to Sankaranarayanan [57]. Due to the matrix crystal characteristics, the first generation of SAW sensors lost a significant portion of their signal inputs. To solve this issue, Sankaranarayanan and his colleagues inserted zinc oxide-filled microcavities that behave like bumpers in a bowling alley by trapping energy near the surface that would otherwise be lost to bulk waves [65,66]. The microcavities, according to Laidoudi et al. [67], cut energy losses by 50%. This means that we are a great deal closer to producing these portable biosensors.

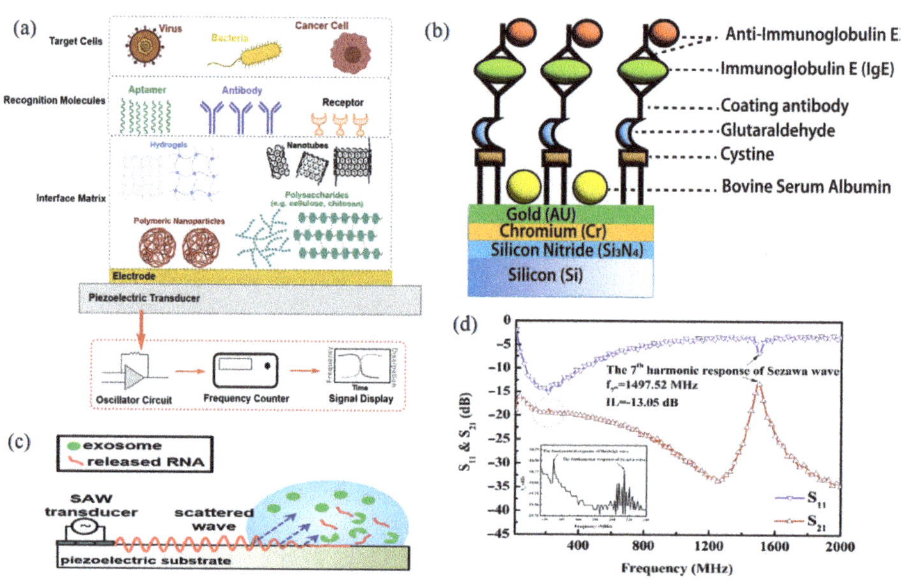

Figure 6. (**a**) Components of an acoustic biosensor to capture whole cells. (**b**) Schematic diagram for the SAW sensor integration of cystamine, glutaraldehyde, and IgE antibody/antigen multilayer. (**c**) Schematic of SAW-based lysing of exosomes to release RNA for detection [24]. (**d**) Frequency response of the Sezawa-mode SAW device [32,68] (copyright permission no: 5543441314248).

6.1. Micromolecular Chemical Analyses by Acoustic Sensors

The natural phytochemical detection by acoustic sensor and phytochemical potential bioactivities for human health led the scientific community to examine the acoustic sensor from a new scientific perspective in terms of their potential uses in facing the demands of the national and international chemical problems such as the hazardous use of the organic solvents and heavy elements. For instance, acoustic sensor products are the least expensive chemical analysis commodity, especially in large-scale applications. Tess and Cox [69] mentioned that the acoustic sensor can serve as a technological source for tracking cellular amino acids. Moreover, the acoustic sensor amin structures have unique functional and technological properties to be used in chemical analysis applications [70].

Researchers have tested several macro- and micromolecular chemical analyses by acoustic sensors combined with other optical techniques to enhance the chemical, functional, and technological properties of chemical analysis and their final collected results. In addition, acoustic sensors can solve traditional technologies such as the dye interference of spectroscopy [1]. As an example, SAW is well known as an excellent choice for deficient protein samples. Recently, several studies recommended the use of a SAW compared to other ASWs due to its pharmaceutical and nutraceutical application high sensitivity due to the increase in the surface area to mass ratio. Therefore, membrane-based systems are promising concerning their sensitivity [71]. In addition, many chemicals were studied using different acoustic intensities of the acoustic transducers which have the potential to select the exact chemical molecule fingerprint [18,72] (Figure 7).

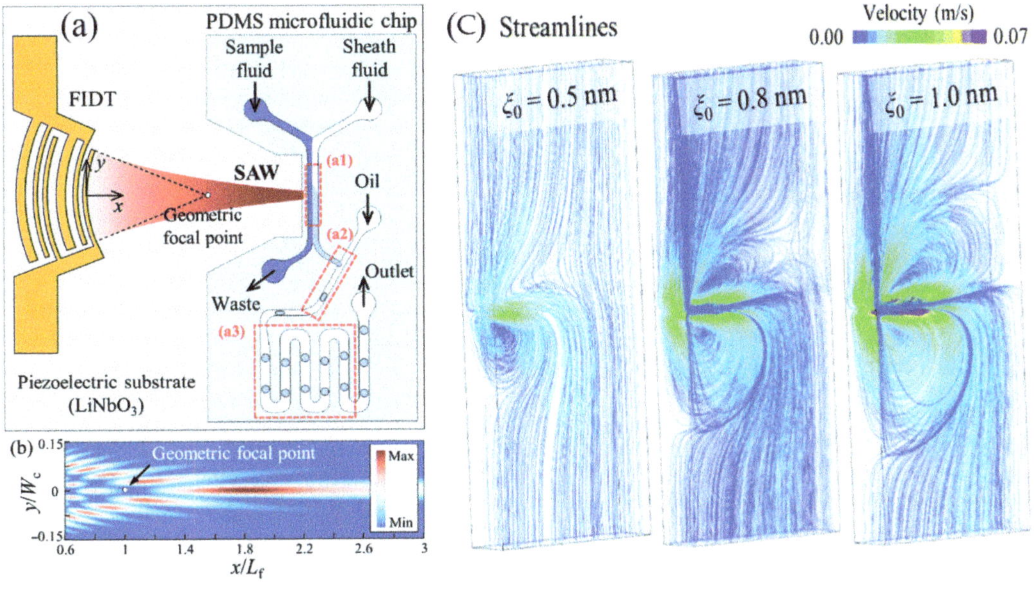

Figure 7. (**a**) Schematic diagram of the acoustofluidic device for the production of droplets with fine-tuned chemical concentrations. (**b**) Numerical simulation of the surface acoustic wave amplitude field for the transducer based on the exact angular spectrum of plane wave theory. (**c**) Variations in the solute concentration with the duty cycle (3.33 to 30%) and the amplitude of the applied voltage (2.23, 2.47, and 2.76 VPP) [18] (copyright permission no: 4859961463671).

Bourdeau et al. [27] developed an acoustic method for in vivo visualizing and imaging of the microbial cellular chemical composition inside the mammalian hosts. This method proved its efficiency in its application as an analytical method. For example, acoustic sensors based on quartz crystal microbalance (QCM) were used to detect tea aroma (e.g., linalool, geraniol, linalool oxide, and Trans-2-hexenal) during its fermentation process [28].

In addition, recent emerging and chemical-free technologies related to in situ detection of the physicochemical changes in the biological media are important for studying the cellular-based levels and the functional components concentrations, such as Raman, circular dichroism (CD), and nuclear magnetic resonance (NMR) [73,74]. Several studies have documented the efficacy of these technologies for the replacement, enhancement, and improvement of various conventional analytical techniques in detecting animal and plant tissues [75–80]. On the other hand, these methods have common linked challenges, such as the fluorescence dye intervention on the used photonic optical sensors. As a proper solution, Garrett and Wang [21] reported that optical acoustic sensors could achieve high levels in biochemical photoacoustic imaging of biological systems. In addition, Westerveld et al. [81] fabricated an optoacoustic imaging system for the mouse brain tissue using >15 MHz acoustical frequency and <100 μm wavelength in water. This tomographic imaging relies on a low detection limit (noise/pressure, NEP); it was mentioned that piezoelectric sensors rely on their mechanical resonance to enhance the signal amplitude. In addition, Tian et al. [82] developed an acoustic phononic crystals method to support the acoustic topological states with complex wavenumbers that can configure the formation of rainbow edge waves of the studied samples. This study may spark future investigations of topological states with complex wavenumbers in the graded materials.

6.2. *Application of Acoustic Sensors in Protein, Lipid and Biomarker Level Detection*

The application of acoustic biosensors as alternative methods for protein and their functional impact analysis has become a popular scientific area [22]. Cho et al. [83] studied the lipid–protein interactions and protein–protein interactions of the cellular membrane by using a QCM biosensor. They reported that QCM could detect the charged zwitterionic functional impact of the tissue lipid bilayer compositions. In addition, Jiang et al. [84] employed acoustic sensors for in vivo imaging of wild plants and in vitro cell imaging using quantum dots technology to bio-image the chemical composition of plants (Figure 8). That could be obtained through the ability of these biosensors for detecting the functional groups of proteins in the biological fluids. For instance, the covalent immobilization of urinary proteins allowed the selective detection of nitroaromatic compounds which may occur in explosives [85]. For instance, Pomowski, Baricham, Rapp, Matern and Lange [30] used SH-SAW coated with 2-methacryloyloxyethyl phosphorylcholine polymer for the label-free detection of the inflammatory marker C-reactive protein in human serum. The authors mentioned that SH-SAW allowed significant differentiation between human CRP serum concentrations lower than 10 mg L^{-1} which could effectively diagnose the bacterial infection. In addition, these sensors have been recently developed for biosensation and detection of SARS-CoV-2-related antibodies. In a study by Peng et al. [86], it was mentioned that SH-SAW achieved high correlation coefficients (R = 0.99) at different concentrations (34.375–1100 ng mL^{-1}) of its protein antibodies, with better sensitivity compared to ELISA.

Acoustic Love waveguide sensors are potential effective tools for detecting the lipid mono- and bilayers through the use of a thiol-coated surface area. The sensitivity of AWS revealed that it could detect the lipid layer mass change during layer deposition and the viscoelastic properties of the interface could change significantly [87].

Regarding the importance of AWS in the biomarker field that is related to the global burden, Onen et al. [88] fabricated a urinary SH-SAW technique for detecting the anti-apoptotic protein of the B-cell lymphoma 2 (Bcl-2) for ovarian cancer early detection. In that study, the sensor was able to successfully detect Bcl-2 in the concentration range of 0.5 to 12 ng mL^{-1}. Thus, it could be applied as a promising technology in the diagnosis and quantification of ovarian cancer. The sensors detect cells mostly via their sensitivity in viscoelasticity and mechanical properties. In particular, the QCM sensor detects cytoskeletal rearrangements caused by specific drugs affecting either actin microfilaments or microtubules.

Acoustic-Based Biosensors for Bio-Imaging the Live Cell Enzymes and Active Ingredients

The use of acoustic sensors for the biochemical analysis of live cells and particles is an emerging technology that integrates acoustics and microfluidics [89]. In the last decade, this technology has attracted significant attention due to its biocompatible, contactless, and label-free nature. For instance, it has been widely validated in the separation of cells, viruses, biomolecules, exosomes, and submicron bioparticles [90].

To increase the detection sensitivity, micro/nano-acoustic biosensors are typically employed to increase the activity of particular biomolecules such as enzymes [91]. These biosensors are built on a special kind of gas vesicles, which are protein nanostructures packed with air that vibrate in response to ultrasound vibrations. For example, the incorporation of acoustic biosensors markedly increased the efficiency of single-cell enzyme activities. This was proved by Lakshmanan et al. [92] who studied the physicochemical characteristics of the bacterial cell-released bioactive enzymes using an acoustic biosensor of endopeptidase. The authors claimed that acoustic sensors at 132 and 477 kPa levels of waves showed high resolution for tracking the single-cell proteases.

Deep tissue can be easily imaged with high spatiotemporal resolution using acoustic waves. In the works of Jiang et al. [84] and Jiang et al. [8], application of quantum dots technology for in vitro cell imaging and in vivo imaging of natural plants allowed for the bio-imaging of plant chemical composition. Moreover, the acoustic biosensors were employed by Lakshmanan, Jin, Nety, Sawyer, Lee-Gosselin, Malounda, Swift, Maresca, and Shapiro [2] to image the activity of enzymes within the mouse gastrointestinal tract. In addition, Barie and Rapp [93] used an XY-cut $LiTaO_3$ SH-wave sensor at 380 MHz acoustic resonance for detecting glucose oxidase (GOD) enzyme. In that study, the authors used antibodies with high specificity that can bind up to 16 ng/mm^2 of protein with a high sensitivity detection limit (59 Hz/ng).

The idea behind using acoustic-based biosensors is to couple the measurement process, such as analyte adsorption, with a change in the acoustic wave's physical characteristics, such as its frequency and velocity, which could be related to the analyte concentration [91]. Because of light scattering and interference with their phytochemicals' fluorescents, existing molecular biosensors based on fluorescent emission are of limited use.

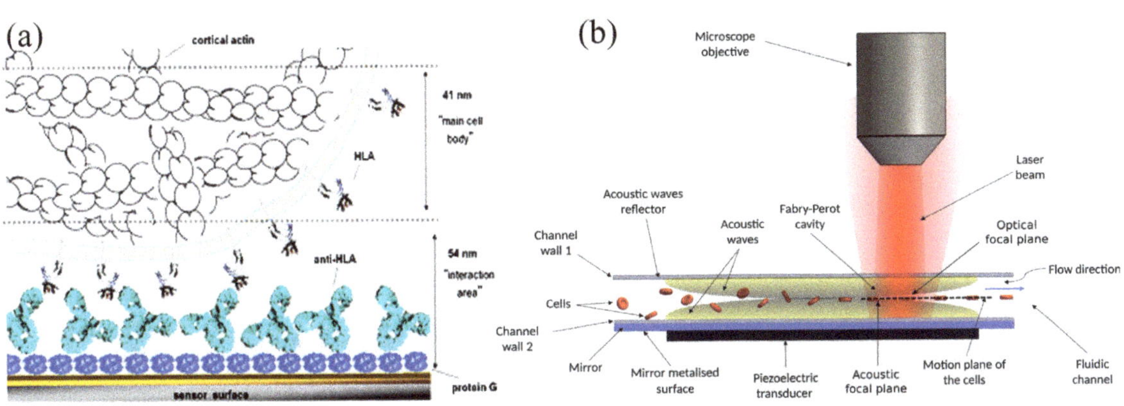

Figure 8. (**a**) Schematic illustration of the acoustic sensor interface for cells interacting via membrane HLA immunoglobulin molecule with surface-immobilized antibodies. The dashed lines for penetration depths of the quartz crystal microbalance (QCM) sensor with 95 nm. (**b**) Microfluidic channel producing acoustic manipulation. (**b**) The cells' microfluidic channel serves as an acoustic reflector sandwich between the wall and the piezoelectric transducer to complete the interferometer through the acoustic waves (yellow) [94,95] (copyright permission no: 600121448).

6.3. Application of Acoustic Sensor Detection of Single Cell Metal Elements

The use of acoustic waves for studying the cellular heavy metal and their interactions with the cell's DNA has been recently established. For example, Beabout et al. [96] used An Echo acoustic transducer for testing the ability of combined sensor reactions to detect multiple heavy metals when the arsenic, cadmium, and mercury DNA circuits are combined into one reaction at a 10 μM limit of detection. In addition, there are several studies on using acoustic sensors for tracking microalgae single-cell elements. Studies showed a strong relationship between acoustic sensor types and their impacts on microalgae chemical composition studies [25,34] (Figure 9). The results of these studies provided an insight into the different interactions of dried biomass of Spirulina (*A. platensis*) with metallic cations using acoustic and microscopic tools. The authors applied an acoustic wave platform to perform real-time monitoring of the interaction of a heavy metal solution in contact with Spirulina cells. Love wave and shear horizontal wave sensors were used, and the authors reported that Love wave sensors were ideally suited for (bio)chemical applications in gases and liquids [40]. In addition, this technique has reached high sensitivity for the characterization of heavy metals at low concentrations (10^{-12} M). Gongi et al. [97] characterize the cadmium (Cd^{2+}) and mercury (Hg^{2+}) heavy metal ions based on the extracellular polymeric substances (EPSs) isolated from a Tunisian thermophilic microalga strain *Graesiella* sp. In that study, the authors used quartz-based AWS and a 1.2×1.2 cm^2 active area with an acoustic resonance of 118 MHz. It was mentioned that Love waves sensors showed good analytical performance and a low detection limit of 10^{-10} M. This could be due to the structural complexity containing hydrophilic and hydrophobic groups of the microalgae EPSs that can absorb and retain the water molecule, which offers them gelling characteristics and increases their ability to interact and adsorb the heavy metals. Additionally, Jiang, Jin and Gui [84] utilized an acoustic-assisted solvothermal process for quantum dot-based bio-imaging of plant zinc ion. The technique's feasibility, according to the authors, might be employed for both in vivo and in vitro imaging of real plants [94,95].

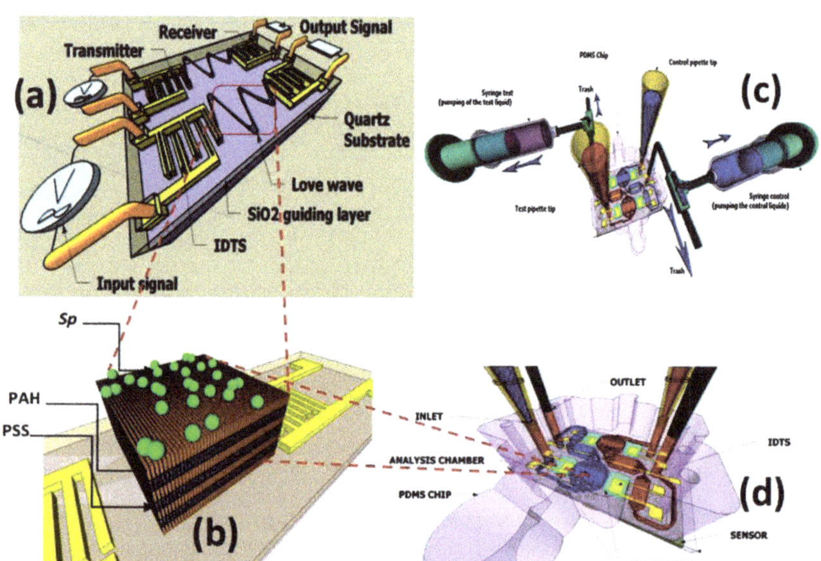

Figure 9. A technological study of the impact of acoustic sensor acoustic sensors on the microalgae chemical composition. Schematic of a SAW sensor with a hybrid biofilm of polyelectrolyte microalgae. (**a**) Scheme of SAW. (**b**) *Spirulina* immobilization on a polyelectrolyte multilayer (PEM) coated with a layer by layer (LBL) method. (**c**,**d**) hydrodynamic chip with microfluidic network, aligned on SAW [25,34] (open access permission).

6.4. Acoustic Based Sensors for Cell-Level Detection

A SAW chemical sensor's reaction time has been demonstrated to be temperature-dependent [98]. The rate at which the analytes diffuse or dissociate may increase as the temperature rises, reducing the response time. For instance, Chen, Chang, Cheng, Shen and Kao [32] reported that the usage of an immobilized SAW sensor with an isolated cavity to measure IgE antibody sensitivity reached 4.44×10^6 cm^2/g. This should have a special condition for the used substrates and conditions for increasing the detection rate of the allergens in the biological fluids and tissues based on the Sezawa acoustic wave frequencies (Figure 6). Furthermore, all sonic wave sensors are sensitive to changes in a wide range of physical characteristics [99]. Many factors influence the rate of layer–analyte interaction and, consequently, the response time of a SAW sensor [100]. In the case of mass-based sensing layers, the rate of diffusion of the adsorbed mass into the film, to the piezoelectric substrate, and back to the film surface heavily influences the response and recovery times of a SAW sensor [101].

It is known that every organism either consists of cells or by itself is a single cell, whether it is an animal, a plant, or a microorganism. Therefore, it is very important to understand the cellular behavior in both biological and biomedical fields by studying the cells' chemical compositions or their interactions in the surrounding environments. For instance, several studies used acoustic sensors for studying cellular activities such as adhesion activity, which is considered one of the most important cell functionality-related biomarkers that are affected by their signaling pathways that direct cell functions [23].

Meanwhile, acoustic biosensors offer the possibility to analyze cell attachment and spreading. This is due to the offered speed of detection, the real-time non-invasive approach, and their high sensitivity not only to mass coupling but also to viscoelastic changes occurring close to the sensor surface. QCM and surface acoustic wave (Love wave) systems have been used to monitor the adhesion of animal cells to various surfaces and record the behavior of cell layers under various conditions [94]. In addition, Mejia Morales, Glynne-Jones, Vassalli and Lippi [95] have established an acoustofuidic interferometric device for high-throughput monitoring analysis of the microalgae cells (*Tetraselmis* sp.) and yeast (*Saccharomyces cerevisiae*) cells based on their physical properties (such as morphology or mechanics). In that method, the authors established a flowing channel with fundamental acoustic mode resonance (6.682 MHz) for passing the cells through it. In addition, the authors identified two types of cell-induced perturbations (strong and weak). The first occurs when the cell crosses the optical resonator's axis, while the second takes place when the cell crosses any other portion of the resonator. As mentioned, that method provided high sensitivity and a speed potentially suitable to obtain the high throughput necessary to handle the variability stemming from the biological diversity of the cells (Figure 8b) [95].

The use of AWS can qualitatively monitor the dynamics and counting the varied cell types by relying on an acoustic wave which penetrated into the basal plane region of the cell. Qualitative data about the cell count and dynamics can be provided based on the used acoustic frequency [102]. A QCM cell composed of a cell culture incubator and a detection system was developed to monitor the growth of epithelial colorectal adenocarcinoma cells (Caco-2). That fabricated system allowed investigations of count for the proliferated and dead cells [103].

6.5. Acoustic-Based Sensors for Monitoring Cell Culture Environment

A fundamental variable in a culture medium such as pH, glucose, CO_2, and sucrose could negatively affect the cells during their cultivation processes. For example, the biological processes are exquisitely sensitive to acid-base chemistry including the concentration of protons H^+ and carbonate (CO_3^{2-}) formation. These conditions should be real-time monitored by advanced sensors for controlling them by an appropriately formulated buffering regime such as CO_2/HCO_3^- [104]. For instance, Wang et al. [105] monitored the pH condition of the tumoroid cultures by SAW sensors. They fabricated novel acoustic pH sensor based on $LiTaO_3$ to generate a 13.91 MHz center frequency for tracking the tumor cell (tumoroid) culture pH (Figure 10). They tracked the pH changes during 5 days of cultivation. The principle of using QCM for pH detection is based on the pH reactive polymer layer to measure the polymer mass loading by shrinkage and swell. This action causes transition function of electrical corrosion between the two oxidation states [106,107].

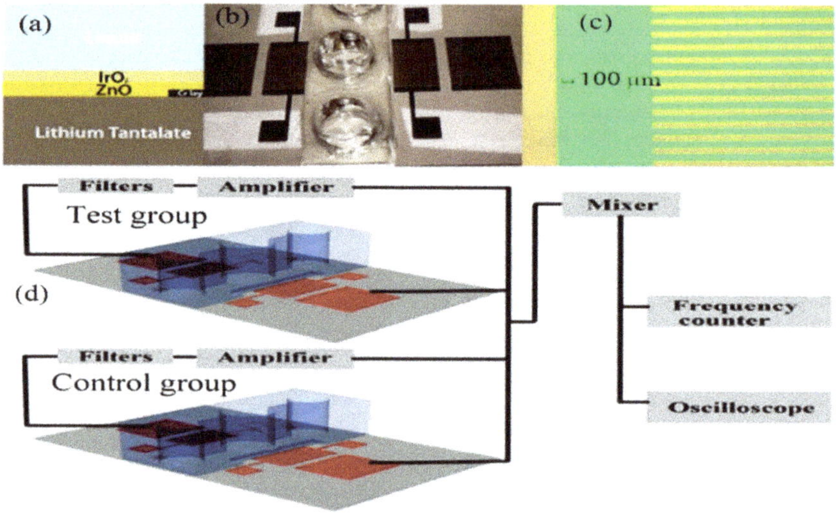

Figure 10. (**a**) Double-layer construction SAW sensor. (**b**) Fabricated resonator and fluidic well. (**c**) Microscopic image of the finger pairs. (**d**) Potential experimental setup illustration [108] (copyright permission: 5567480548369).

6.6. Acoustic Sensors for Cancer and Tumor-Level Detection and Tumoroid Cultures

The application of acoustic sensors, especially SAW, for differentiating the healthy cells from the aggressive and nonaggressive tumor cells and monitoring the cancer cell activities has emerged as a new significant technology [102]. This technology is among the key technologies for tumor and cancer biomarker detection, in which its frequency sensitivity could reach 8.704 pg/Hz with a mass sensitivity of 2810.25 m^2/kg for detecting mammoglobin cancer-related antigens [109]. Zhang et al. [110] used a recyclable chitosan-based QCM biosensor for real-time detection of breast cancer cells. The authors compared human erythrocyte, endothelial cell, and oral epithelial cells by using a 5 MHz acoustic resonance. The obtained results showed that QCM successfully discriminated the different cell types with a wide linear range of 4.5×10^2–1.01×10^5 cells/mL and a detection limit of 430 cells/mL. Hianik [108] reported the effectiveness of the fabricated immunoacoustic biosensor for diagnostics of leukemia blood cell cancer. For that purpose, the sensor operates at frequencies of around 100 MHz. In another study by Hao et al. [111], the authors established an immunoacoustic biosensor for diagnostics of leukemia (blood cell cancer). That applied study used SH-SAW for the detection and separation of Jurkat and K562 leukemic cells by using a 122.5 MHz frequency. The authors mentioned that maximal sensitivity of detection was 10^3 Jurkat cells/mL after a 15 min detection time. That method successfully differentiated the cells in a mixture of Jurkat/K562 cells (1:1000) at a 10^6 cells/mL concentration. Meanwhile, SAW has been used in the field of tumoroid culture detection. Wang et al. [112] designed a SAW system for drug sensitivity assay of the colorectal perfused tumoroid cultures (Figure 11). That study was focused on the tumoroid cell proliferation, density and pH. During the experiments, the wells mentioned in Figure 11 were filled with the tumor cells and culture medium, and the control group was filled with blank culture medium. The principal theory, when the SAWs propagate through the detection area where the nanofiber scaffold was attached, the phase velocity was changed due to the mass loading changes caused by tumor cell growth. As the tumoroids grow, the frequency and phase for the test group sensor changes while the frequency of the control group sensor remains nearly constant. In addition, when the acoustic waves travel through the well that has pH electrodes, the potential difference is generated between the two electrodes located on the bottom and top of the two sensor layers. The potential difference is affected by the conductivity and dielectric properties of the liquid and electrode layers. As a result, an impulse signal to the input interdigital transducers was generated that could distinguish the cell densities and the pH calculations.

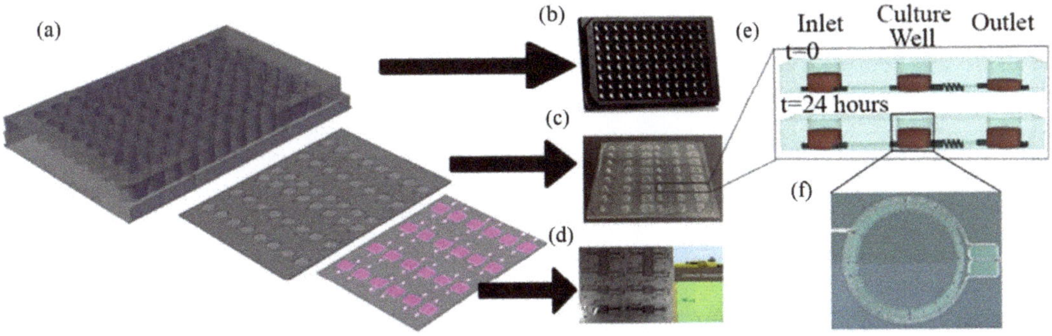

Figure 11. Design of the SAW-based gravitational microfluidic system. (**a**) Microfluidic 96-well plate with integrated sensor layer at the bottom. (**b**) Bottomless 96-well plate. (**c**) Microfluidic channel layer. (**d**) Surface-acoustic-wave (SAW)-based sensor layer. (**e**) 3D vertical image of the well fluid levels at 0 h and after 24 h. (**f**) A real image horizontal view of the culture well (copyright permission: 5567500483871).

6.7. Application of Acoustic Sensors in Biological Fluid-Level Detection

The AWS has the ability to detect chemicals of the biological fluids and liquids. This could be due to the fact that the AWS device waves have poor performance in liquids as the propagating wave's vertical component is blocked by the liquid. Additionally, for liquid sensing, the Love wave acoustic sensor has the maximum sensitivity for this purpose [113]. Vellekoop [114] reported that the ST-quartz shear horizontal (SH) acoustic plate mode (APM) sensor can distribute throughout the bulk of the substrate that is developed for sensing liquid particle displacements. The shear wave transmits in the low shear acoustic velocity material (upper layer) with a higher shear acoustic velocity, thus representing bilayer geometry. At a specific frequency, the Love wave device provides huge design suppleness, where the energy limitation is determined by deposited overlayer thickness and the acoustic properties. In the liquid sample, within about 60 nm from the device surface, the evanescent field of the shear acoustic wave probes' electric, viscosity, and mass changes occurs (Figure 8b). Consequently, by monitoring the acoustic wave propagation characteristics, including the frequency, phase, and amplitude, it is probable to detect the binding kinetics and obtain the corresponding acoustic to the optical immuno-sensor [89].

6.8. Application of Acoustic Sensors in Tissue-Level Detection

According to Garrett and Wang [21], the acoustic sensors have significant sensitivity for tight focus inside biological tissues. As an example, the authors mentioned that a transducer with 100 MHz and an active area = 30 mm^2 can achieve a 0.06–0.6 mPa Hz^{-1} sensitivity. It could virtually detect the chemical composition inside the biological tissues by 0.7λ (λ: acoustic wavelength). Lakshmanan, Jin, Nety, Sawyer, Lee-Gosselin, Malounda, Swift, Maresca and Shapiro [92] used acoustic biosensors for easy bioimaging of the chemicals of live tissues [115]. In that study, they used 1.2×1.2 mm^2 of C57BL/6J male mice colon lumen tissues for tracking the endopeptidase enzyme concentrations that were produced by *E. coli* in the deep animal tissues with high spatiotemporal resolution (below 100 µm and 1 ms, respectively). The tissues had acoustic impedance values resulting in relatively strong reflections; the biological tissues and the wide range of their fibers, cells, and organelles affects the acoustic wave scatters. As an example, Jathoul et al. [116] developed a photoacoustic imaging system that allowed in vivo high-resolution imaging at a depth beyond that of the normal optical microscopy (10 mm in depth with a spatial resolution below 100 μm). More importantly, that system discriminated the tyrosinase enzyme composition and its genetic expressing in the live tissues of the non-vascularized invisible tissues with selective labeling of their cells. Each tissue is characterized by a different attenuation coefficient value, which increases nonlinearly with frequency [117]. As the acoustic waves pass through the tissues, the acoustic waves also deposit momentum into that tissue, resulting in mechanical forces known as acoustic radiation forces (ARF) which could be easily detected and analyzed [118].

7. Conclusions

Acoustic wave sensors are extremely versatile devices that are just beginning to realize their commercial potential. Meanwhile, the ASW sensors require no high operating power and preparation for monitoring of real-time cellular chemical, morphological, and functional characteristics. Other applications include measuring cell acceleration, real-time shock, viscosity, and biological media composition. It is highly evident from the above presentation that ASW techniques are in high demand and an active research area because of their potential application to different biological and chemical applications as biosensors. Even though there are a wide variety of applications of these devices with several advantages and disadvantages, there are many opportunities where improvements can be made for future applications of these sensitive devices. In the field of microfluidics, several biosensors and other chemical sensors have been used to facilitate the access to rapid and cost-effective diagnostic platforms. In addition, these sensors have an acoustoelectric sensitivity, allowing the detection of low concentrations of toxins, heavy metals, and biomarker proteins. Love, Sezawa, and Rayleigh wave sensors have been proven to be the most sensitive in general as a result of their larger energy density on the surface of the studied solid and liquid samples. Therefore, much work is continuing in developing these important sensors for their high-accuracy future applications.

Author Contributions: Conceptualization, M.G.; validation, M.G., H.S.G., N.A., X.L.; writing—original draft preparation, M.G.; writing—review and editing, M.G., H.S.G., N.A., X.L.; project administration, M.G., H.S.G., N.A., X.L.; funding acquisition, M.G., H.S.G., N.A. All authors have read and agreed to the published version of the manuscript.

Funding: This work was funded by the "Belt and Road" joint project fund between Zhejiang University, China, and National Research Centre, Egypt (Project No: SQ2023YFE0103360). It was also funded by the Deputyship for Research Innovation, Ministry of Education in Saudi Arabia for funding this research work through the project number (Grant No. INSTV007). Additionally, this research work was supported by the National Natural Science Foundation of China (Project No: 32171889) and the Key R&D Projects in Zhejiang Province (Project No: 2022C02044, 2023C02009, 2023C02043).

Institutional Review Board Statement: Not applicable.

Informed Consent Statement: Not applicable.

Data Availability Statement: Not applicable.

Acknowledgments: The authors would like to thank MDPI for facilitating the launch of this Special Issue. The authors extend their appreciation to the Deputyship for Research Innovation, Ministry of Education in Saudi Arabia for funding this research work through the project number (Grant No. INSTV007).

Conflicts of Interest: The authors declare no conflict of interest.

Sample Availability: Not applicable.

References

1. Gouda, M.; Bekhit, A.E.D.; Tang, Y.; Huang, Y.; Huang, L.; He, Y.; Li, X. Recent innovations of ultrasound green technology in herbal phytochemistry: A review. *Ultrason. Sonochemistry* **2021**, *73*, 105538. [CrossRef] [PubMed]
2. Gouda, M.; He, Y.; Bekhit, A.E.-D.; Li, X. Emerging Technologies for Detecting the Chemical Composition of Plant and Animal Tissues and Their Bioactivities: An Editorial. *Molecules* **2022**, *27*, 2620. [CrossRef] [PubMed]
3. Kuznetsova, I.E.; Anisimkin, V.I.; Gubin, S.P.; Tkachev, S.V.; Kolesov, V.V.; Kashin, V.V.; Zaitsev, B.D.; Shikhabudinov, A.M.; Verona, E.; Sun, S. Super high sensitive plate acoustic wave humidity sensor based on graphene oxide film. *Ultrasonics* **2017**, *81*, 135–139. [CrossRef]
4. De Jong, M.; Chen, W.; Geerlings, H.; Asta, M.; Persson, K.A. A database to enable discovery and design of piezoelectric materials. *Sci. Data* **2015**, *2*, 150053. [CrossRef] [PubMed]
5. Li, C.; Kan, H.; Luo, J.; Fu, C.; Zhou, J.; Liu, X.; Wang, W.; Wei, Q.; Fu, Y. A high performance surface acoustic wave visible light sensor using novel materials: Bi_2S_3 nanobelts. *RSC Adv.* **2020**, *10*, 8936–8940. [CrossRef]
6. Rana, L.; Gupta, R.; Tomar, M.; Gupta, V. ZnO/ST-Quartz SAW resonator: An efficient NO_2 gas sensor. *Sens. Actuators B Chem.* **2017**, *252*, 840–845. [CrossRef]
7. Grate, J.W.; Martin, S.J.; White, R.M. Acoustic Wave Microsensors. *Anal. Chem.* **2012**, *65*, 940A–948A. [CrossRef]
8. Long, G.; Guo, Y.; Li, W.; Tang, Q.; Zu, X.; Ma, J.; Du, B.; Fu, Y. Surface acoustic wave ammonia sensor based on ZnS mucosal-like nanostructures. *Microelectron. Eng.* **2020**, *222*, 111201. [CrossRef]
9. Wohltjen, H.; Dessy, R. Surface acoustic wave probes for chemical analysis. III. Thermomechanical polymer analyzer. *Anal. Chem.* **2002**, *51*, 1470–1475. [CrossRef]
10. Liu, X.; Chen, X.; Yang, Z.; Xia, H.; Zhang, C.; Wei, X. Surface acoustic wave based microfluidic devices for biological applications. *Sens. Diagn.* **2023**, *2*, 507–528. [CrossRef]
11. Zhang, Y.; Cai, Y.; Zhou, J.; Xie, Y.; Xu, Q.; Zou, Y.; Guo, S.; Xu, H.; Sun, C.; Liu, S. Surface acoustic wave-based ultraviolet photodetectors: A review. *Sci. Bull.* **2020**, *65*, 587–600. [CrossRef] [PubMed]
12. Viggen, E.M.; Arnestad, H.K. Modelling acoustic radiation from vibrating surfaces around coincidence: Radiation into fluids. *J. Sound Vib.* **2023**, *560*, 117787. [CrossRef]
13. Witte, M.; Paszkiewicz, A.; Ospel, M.W.; Rathje, J.T.; Hieke, M.; Wurm, F.-H. Design of a hydro sound intensity probe for quantification and localization of acoustic sources—Applied to a hubless marine rim drive. *Ocean. Eng.* **2023**, *267*, 113227. [CrossRef]
14. Zou, C.; Harne, R.L. Deployable tessellated transducer array for ultrasound focusing and bio-heat generation in a multilayer environment. *Ultrasonics* **2020**, *104*, 106108. [CrossRef]
15. Li, J.; Piwakowski, B. Time domain model and experimental validation of non-contact surface wave scanner. *Ultrasonics* **2019**, *94*, 242–263. [CrossRef]
16. Singer, E.A.; Golijanin, D.J.; Davis, R.S.; Dogra, V. What's new in urologic ultrasound? *Urol. Clin. N. Am.* **2006**, *33*, 279–286. [CrossRef]
17. Lange, K. Bulk and Surface Acoustic Wave Sensor Arrays for Multi-Analyte Detection: A Review. *Sensors* **2019**, *19*, 5382. [CrossRef]
18. Park, J.; Destgeer, G.; Afzal, M.; Sung, H.J. Acoustofluidic generation of droplets with tunable chemical concentrations. *Lab Chip* **2020**, *20*, 3922–3929. [CrossRef]
19. Ali, W.R.; Prasad, M. Piezoelectric MEMS based acoustic sensors: A review. *Sens. Actuators A Phys.* **2020**, *301*, 111756. [CrossRef]
20. Wu, Y.; Ma, Y.; Zheng, H.; Ramakrishna, S. Piezoelectric materials for flexible and wearable electronics: A review. *Mater. Des.* **2021**, *211*, 110164. [CrossRef]
21. Garrett, D.C.; Wang, L.V. Acoustic sensing with light. *Nat. Photonics* **2021**, *15*, 324–326. [CrossRef]
22. Matatagui, D.; Fontecha, J.; Fernández, M.J.; Oliver, M.J.; Hernando-García, J.; Sánchez-Rojas, J.L.; Gràcia, I.; Cané, C.; Santos, J.P.; Horrillo, M.C. Comparison of two types of acoustic biosensors to detect immunoreactions: Love-wave sensor working in dynamic mode and QCM working in static mode. *Sens. Actuators B Chem.* **2013**, *189*, 123–129. [CrossRef]
23. Wu, H.; Zu, H.; Wang, J.H.; Wang, Q.M. A study of Love wave acoustic biosensors monitoring the adhesion process of tendon stem cells (TSCs). *Eur. Biophys. J.* **2019**, *48*, 249–260. [CrossRef] [PubMed]
24. Taller, D.; Richards, K.; Slouka, Z.; Senapati, S.; Hill, R.; Go, D.B.; Chang, H.C. On-chip surface acoustic wave lysis and ion-exchange nanomembrane detection of exosomal RNA for pancreatic cancer study and diagnosis. *Lab Chip* **2015**, *15*, 1656–1666. [CrossRef]
25. Tekaya, N.; Gammoudi, I.; Braiek, M.; Tarbague, H.; Moroté, F.; Raimbault, V.; Sakly, N.; Rebière, D.; Ben Ouada, H.; Lagarde, F.; et al. Acoustic, electrochemical and microscopic characterization of interaction of Arthrospira platensis biofilm and heavy metal ions. *J. Environ. Chem. Eng.* **2013**, *1*, 609–619. [CrossRef]
26. Moll, N.; Pascal, E.; Dinh, D.H.; Pillot, J.P.; Bennetau, B.; Rebiere, D.; Moynet, D.; Mas, Y.; Mossalayi, D.; Pistre, J.; et al. A Love wave immunosensor for whole *E. coli* bacteria detection using an innovative two-step immobilisation approach. *Biosens. Bioelectron.* **2007**, *22*, 2145–2150. [CrossRef]
27. Bourdeau, R.W.; Lee-Gosselin, A.; Lakshmanan, A.; Farhadi, A.; Kumar, S.R.; Nety, S.P.; Shapiro, M.G. Acoustic reporter genes for noninvasive imaging of microorganisms in mammalian hosts. *Nature* **2018**, *553*, 86–90. [CrossRef]

28. Sharma, P.; Ghosh, A.; Tudu, B.; Sabhapondit, S.; Baruah, B.D.; Tamuly, P.; Bhattacharyya, N.; Bandyopadhyay, R. Monitoring the fermentation process of black tea using QCM sensor based electronic nose. *Sens. Actuators B Chem.* **2015**, *219*, 146–157. [CrossRef]
29. Di Pietrantonio, F.; Benetti, M.; Cannata, D.; Verona, E.; Palla-Papavlu, A.; Fernandez-Pradas, J.M.; Serra, P.; Staiano, M.; Varriale, A.; D'Auria, S. A surface acoustic wave bio-electronic nose for detection of volatile odorant molecules. *Biosens. Bioelectron.* **2015**, *67*, 516–523. [CrossRef]
30. Pomowski, A.; Baricham, C.; Rapp, B.E.; Matern, A.; Lange, K. Acoustic Biosensors Coated With Phosphorylcholine Groups for Label-Free Detection of Human C-Reactive Protein in Serum. *IEEE Sens. J.* **2015**, *15*, 4388–4392. [CrossRef]
31. Agostini, M.; Greco, G.; Cecchini, M. A Rayleigh surface acoustic wave (R-SAW) resonator biosensor based on positive and negative reflectors with sub-nanomolar limit of detection. *Sens. Actuators B Chem.* **2018**, *254*, 1–7. [CrossRef]
32. Chen, Y.-C.; Chang, W.-T.; Cheng, C.-C.; Shen, J.-Y.; Kao, K.-S. Development of human IgE biosensor using Sezawa-mode SAW devices. *Curr. Appl. Phys.* **2014**, *14*, 608–613. [CrossRef]
33. Du, J.; Harding, G.L.; Ogilvy, J.A.; Dencher, P.R.; Lake, M. A study of Love-wave acoustic sensors. *Sens. Actuators A Phys.* **1996**, *56*, 211–219. [CrossRef]
34. Gouda, M.; Tadda, M.A.; Zhao, Y.; Farmanullah, F.; Chu, B.; Li, X.; He, Y. Microalgae Bioactive Carbohydrates as a Novel Sustainable and Eco-Friendly Source of Prebiotics: Emerging Health Functionality and Recent Technologies for Extraction and Detection. *Front. Nutr.* **2022**, *9*, 806692. [CrossRef]
35. Hickernell, F.S. Shear horizontal BG surface acoustic waves on piezoelectrics: A historical note. *IEEE Trans. Ultrason. Ferroelectr. Freq. Control.* **2005**, *52*, 809–811. [CrossRef]
36. Ji, J.; Pang, Y.; Li, D.; Huang, Z.; Zhang, Z.; Xue, N.; Xu, Y.; Mu, X. An aptamer-based shear horizontal surface acoustic wave biosensor with a CVD-grown single-layered graphene film for high-sensitivity detection of a label-free endotoxin. *Microsyst. Nanoeng.* **2020**, *6*, 4. [CrossRef]
37. Burtin, A.; Hovius, N.; Turowski, J.M. Seismic monitoring of torrential and fluvial processes. *Earth Surf. Dyn.* **2016**, *4*, 285–307. [CrossRef]
38. Rayleigh, L. On Waves Propagated along the Plane Surface of an Elastic Solid. *Proc. Lond. Math. Soc.* **1885**, *s1-17*, 4–11. [CrossRef]
39. Wang, L. Metal-organic frameworks for QCM-based gas sensors: A review. *Sens. Actuators A Phys.* **2020**, *307*, 111984. [CrossRef]
40. Newton, M.I.; McHale, G.; Martin, F. Experimental study of Love wave devices with thick guiding layers. *Sens. Actuators A Phys.* **2004**, *109*, 180–185. [CrossRef]
41. Hadj-Larbi, F.; Serhane, R. Sezawa SAW devices: Review of numerical-experimental studies and recent applications. *Sens. Actuators A Phys.* **2019**, *292*, 169–197. [CrossRef]
42. Zhang, H.; Wang, H. Investigation of Surface Acoustic Wave Propagation Characteristics in New Multilayer Structure: SiO_2/IDT/$LiNbO_3$/Diamond/Si. *Micromachines* **2021**, *12*, 1286. [CrossRef] [PubMed]
43. Kuznetsova, I.E.; Anisimkin, V.I.; Kolesov, V.V.; Kashin, V.V.; Osipenko, V.A.; Gubin, S.P.; Tkachev, S.V.; Verona, E.; Sun, S.; Kuznetsova, A.S. Sezawa wave acoustic humidity sensor based on graphene oxide sensitive film with enhanced sensitivity. *Sens. Actuators B Chem.* **2018**, *272*, 236–242. [CrossRef]
44. Suenaga, R.; Suzuki, M.; Kakio, S.; Ohashi, Y.; Arakawa, M.; Kushibiki, J.-I. Propagation properties of leaky surface acoustic wave on water-loaded piezoelectric substrate. *Jpn. J. Appl. Phys.* **2018**, *57*, 07LC10. [CrossRef]
45. Zou, Y.; Gao, C.; Zhou, J.; Liu, Y.; Xu, Q.; Qu, Y.; Liu, W.; Soon, J.B.W.; Cai, Y.; Sun, C. Aluminum scandium nitride thin-film bulk acoustic resonators for 5G wideband applications. *Microsyst. Nanoeng.* **2022**, *8*, 124. [CrossRef]
46. Kumar, A.; Prajesh, R. The potential of acoustic wave devices for gas sensing applications. *Sens. Actuators A Phys.* **2022**, *339*, 113498. [CrossRef]
47. Gomes, M.T. Bulk Acoustic Wave Sensors in Chemical Analysis. In *Smart Sensors and MEMS*; Yurish, S.Y., Gomes, M.T., Eds.; NATO Science Series; Springer: Dordrecht, The Netherlands, 2004; Volume 181.
48. Zhang, Y.; Luo, J.; Flewitt, A.J.; Cai, Z.; Zhao, X. Film bulk acoustic resonators (FBARs) as biosensors: A review. *Biosens. Bioelectron.* **2018**, *116*, 1–15. [CrossRef]
49. Durukan, Y.; Shevelko, M.; Peregudov, A.; Popkova, E.; Shevchenko, S. The Effect of a Rotating Medium on Bulk Acoustic Wave Polarization: From Theoretical Considerations to Perspective Angular Motion Sensor Design. *Sensors* **2020**, *20*, 2487. [CrossRef]
50. Gouda, M.; Nassarawa, S.S.; Gupta, S.D.; Sanusi, N.I.; Nasiru, M.M. Evaluation of carbon dioxide elevation on phenolic compounds and antioxidant activity of red onion (*Allium cepa* L.) during postharvest storage. *Plant Phys. Biochem.* **2023**, *200*, 107752. [CrossRef]
51. Mujahid, A.; Dickert, F. Surface Acoustic Wave (SAW) for Chemical Sensing Applications of Recognition Layers. *Sensors* **2017**, *17*, 2716. [CrossRef] [PubMed]
52. Casalinuovo, I.A.; Pierro, D.; Bruno, E.; Francesco, P.; Coletta, M. Experimental use of a new surface acoustic wave sensor for the rapid identification of bacteria and yeasts. *Lett. Appl. Microbiol.* **2006**, *42*, 24–29. [CrossRef] [PubMed]
53. Kiontke, A.; Roudini, M.; Billig, S.; Fakhfouri, A.; Winkler, A.; Birkemeyer, C. Author Correction: Surface acoustic wave nebulization improves compound selectivity of low-temperature plasma ionization for mass spectrometry. *Sci. Rep.* **2021**, *11*, 11620. [CrossRef] [PubMed]
54. Devkota, J.; Ohodnicki, P.R.; Greve, D.W. SAW Sensors for Chemical Vapors and Gases. *Sensors* **2017**, *17*, 801. [CrossRef] [PubMed]
55. Khodagholy, D.; Malliaras, G.G.; Owens, R.M. 8.05-Polymer-Based Sensors. In *Polymer Science: A Comprehensive Reference*; Elsevier: Alpharetta, GA, USA, 2012; Volume 8, pp. 101–128.

56. Gronewold, T.M. Surface acoustic wave sensors in the bioanalytical field: Recent trends and challenges. *Anal. Chim. Acta* **2007**, *603*, 119–128. [CrossRef]
57. Sankaranarayanan, S.K.R.S.; Bhethanabotla, V.R. Design of efficient focused surface acoustic wave devices for potential microfluidic applications. *J. Appl. Phys.* **2008**, *103*, 064518. [CrossRef]
58. Aleksandrova, M.; Badarov, D. Recent Progress in the Topologies of the Surface Acoustic Wave Sensors and the Corresponding Electronic Processing Circuits. *Sensors* **2022**, *22*, 4917. [CrossRef]
59. Hsu, J.-C.; Chao, C.-L. Full-wave modeling of micro-acoustofluidic devices driven by standing surface acoustic waves for microparticle acoustophoresis. *J. Appl. Phys.* **2020**, *128*, 124502. [CrossRef]
60. Jiang, Y.; Tan, C.Y.; Tan, S.Y.; Wong, M.S.F.; Chen, Y.F.; Zhang, L.; Yao, K.; Gan, S.K.E.; Verma, C.; Tan, Y.-J. SAW sensor for Influenza A virus detection enabled with efficient surface functionalization. *Sens. Actuators B Chem.* **2015**, *209*, 78–84. [CrossRef]
61. Lange, K.; Gruhl, F.J.; Rapp, M. Surface Acoustic Wave (SAW) biosensors: Coupling of sensing layers and measurement. *Microfluid. Diagn.* **2013**, *949*, 491–505. [CrossRef]
62. Wang, J.L.; Guo, Y.J.; Li, D.J.; Long, G.D.; Tang, Q.B.; Zu, X.T.; Ma, J.Y.; Du, B.; Tang, Y.L.; Torun, H.; et al. Bacterial cellulose coated ST-cut quartz surface acoustic wave humidity sensor with high sensitivity, fast response and recovery. *Smart Mater. Struct.* **2020**, *29*, 045037. [CrossRef]
63. Wang, X.; Du, L.; Cheng, L.; Zhai, S.; Zhang, C.; Wang, W.; Liang, Y.; Yang, D.; Chen, Q.; Lei, G. Pd/Ni nanowire film coated SAW hydrogen sensor with fast response. *Sens. Actuators B Chem.* **2022**, *351*, 130952. [CrossRef]
64. Mandal, D.; Banerjee, S. Surface Acoustic Wave (SAW) Sensors: Physics, Materials, and Applications. *Sensors* **2022**, *22*, 820. [CrossRef]
65. Pandey, R.K.; Dutta, J.; Brahma, S.; Rao, B.; Liu, C.A.P. Review on ZnO-based piezotronics and piezoelectric nanogenerators: Aspects of piezopotential and screening effect. *J. Phys. Mater.* **2021**, *4*, 044011. [CrossRef]
66. Pan, C.; Han, Y.; Lu, J. Design and Optimization of Lattice Structures: A Review. *Appl. Sci.* **2020**, *10*, 6374. [CrossRef]
67. Laidoudi, F.; Amara, S.; Caliendo, C.; Boubenider, F.; Kanouni, F.; Assali, A. High quality and low loss surface acoustic wave SAW resonator based on chromium-doped AlN on sapphire. *Appl. Phys. A* **2021**, *127*, 255. [CrossRef]
68. Damiati, S. *Acoustic Biosensors for Cell Research*; Springer: Cham, Switzerland, 2020.
69. Tess, M.E.; Cox, J.A. Chemical and biochemical sensors based on advances in materials chemistry. *J. Pharm. Biomed. Anal.* **1999**, *19*, 55–68. [CrossRef]
70. Peng, H.; Liang, C.; Zhou, A.; Zhang, Y.; Xie, Q.; Yao, S. Development of a new atropine sulfate bulk acoustic wave sensor based on a molecularly imprinted electrosynthesized copolymer of aniline with o-phenylenediamine. *Anal. Chim. Acta* **2000**, *423*, 221–228. [CrossRef]
71. Valentine, J.E.; Przybycien, T.M.; Hauan, S. Design of acoustic wave biochemical sensors using micro-electro-mechanical systems. *J. Appl. Phys.* **2007**, *101*, 064508. [CrossRef]
72. Betteridge, D.; Joslin, M.T.; Lilley, T. Acoustic emissions from chemical reactions. *Anal. Chem.* **2002**, *53*, 1064–1073. [CrossRef]
73. Gouda, M.; Chen, K.; Li, X.; Liu, Y.; He, Y. Detection of microalgae single-cell antioxidant and electrochemical potentials by gold microelectrode and Raman micro-spectroscopy combined with chemometrics. *Sens. Actuators B Chem.* **2021**, *329*, 129229. [CrossRef]
74. Gouda, M.; Huang, Z.; Liu, Y.; He, Y.; Li, X. Physicochemical impact of bioactive terpenes on the microalgae biomass structural characteristics. *Bioresour. Technol.* **2021**, *334*, 125232. [CrossRef] [PubMed]
75. Chu, H.; Zhang, C.; Wang, M.; Gouda, M.; Wei, X.; He, Y.; Liu, Y. Hyperspectral imaging with shallow convolutional neural networks (SCNN) predicts the early herbicide stress in wheat cultivars. *J. Hazard. Mater.* **2022**, *421*, 126706. [CrossRef]
76. Zhao, Y.; Zhang, J.; Gouda, M.; Zhang, C.; Lin, L.; Nie, P.; Ye, H.; Huang, W.; Ye, Y.; Zhou, C.; et al. Structure analysis and non-invasive detection of cadmium-phytochelatin2 complexes in plant by deep learning Raman spectrum. *J. Hazard. Mater.* **2022**, *427*, 128152. [CrossRef]
77. Rehman, K.u.; Gouda, M.; Zaman, U.; Tahir, K.; Khan, S.U.; Saeed, S.; Khojah, E.; El-Beltagy, A.; Zaky, A.A.; Naeem, M.; et al. Optimization of Platinum Nanoparticles (PtNPs) Synthesis by Acid Phosphatase Mediated Eco-Benign Combined with Photocatalytic and Bioactivity Assessments. *Nanomaterials* **2022**, *12*, 1079. [CrossRef]
78. Lv, J.-M.; Gouda, M.; El-Din Bekhit, A.; He, Y.-K.; Ye, X.-Q.; Chen, J.-C. Identification of novel bioactive proanthocyanidins with potent antioxidant and anti-proliferative activities from kiwifruit leaves. *Food Biosci.* **2022**, *46*, 101554. [CrossRef]
79. Taha, M.F.; Abdalla, A.; ElMasry, G.; Gouda, M.; Zhou, L.; Zhao, N.; Liang, N.; Niu, Z.; Hassanein, A.; Al-Rejaie, S.; et al. Using Deep Convolutional Neural Network for Image-Based Diagnosis of Nutrient Deficiencies in Plants Grown in Aquaponics. *Chemosensors* **2022**, *10*, 45. [CrossRef]
80. Zong, W.; Gouda, M.; Cai, E.; Wang, R.; Xu, W.; Wu, Y.; Munekata, P.E.S.; Lorenzo, J.M. The Antioxidant Phytochemical Schisandrin A Promotes Neural Cell Proliferation and Differentiation after Ischemic Brain Injury. *Molecules* **2021**, *26*, 7466. [CrossRef] [PubMed]
81. Westerveld, W.J.; Mahmud-Ul-Hasan, M.; Shnaiderman, R.; Ntziachristos, V.; Rottenberg, X.; Severi, S.; Rochus, V. Sensitive, small, broadband and scalable optomechanical ultrasound sensor in silicon photonics. *Nat. Photonics* **2021**, *15*, 341–345. [CrossRef]
82. Tian, Z.; Shen, C.; Li, J.; Reit, E.; Bachman, H.; Socolar, J.E.S.; Cummer, S.A.; Jun Huang, T. Dispersion tuning and route reconfiguration of acoustic waves in valley topological phononic crystals. *Nat. Commun.* **2020**, *11*, 762. [CrossRef]

83. Cho, N.J.; Frank, C.W.; Kasemo, B.; Hook, F. Quartz crystal microbalance with dissipation monitoring of supported lipid bilayers on various substrates. *Nat. Protoc.* **2010**, *5*, 1096–1106. [CrossRef]
84. Jiang, X.; Jin, H.; Gui, R. Visual bio-detection and versatile bio-imaging of zinc-ion-coordinated black phosphorus quantum dots with improved stability and bright fluorescence. *Biosens. Bioelectron.* **2020**, *165*, 112390. [CrossRef]
85. Scorsone, E.; Manai, R.; Ricatti, M.J.; Redaelli, M.; Bergonzo, P.; Persaud, K.C.; Mucignat, C. Major Urinary Proteins on Nanodiamond-Based Resonators Toward Artificial Olfaction. *IEEE Sens. J.* **2016**, *16*, 6543–6550. [CrossRef]
86. Peng, Y.C.; Cheng, C.H.; Yatsuda, H.; Liu, S.H.; Liu, S.J.; Kogai, T.; Kuo, C.Y.; Wang, R.Y.L. A Novel Rapid Test to Detect Anti-SARS-CoV-2 N Protein IgG Based on Shear Horizontal Surface Acoustic Wave (SH-SAW). *Diagnostics* **2021**, *11*, 1838. [CrossRef]
87. Gizeli, E.; Lowe, C.R.; Liley, M.; Vogel, H. Detection of supported lipid layers with the acoustic Love waveguide device: Application to biosensors. *Sens. Actuators B Chem.* **1996**, *34*, 295–300. [CrossRef]
88. Onen, O.; Sisman, A.; Gallant, N.D.; Kruk, P.; Guldiken, R. A urinary Bcl-2 surface acoustic wave biosensor for early ovarian cancer detection. *Sensors* **2012**, *12*, 7423–7437. [CrossRef]
89. Fan, Y.; Wang, X.; Ren, J.; Lin, F.; Wu, J. Recent advances in acoustofluidic separation technology in biology. *Microsyst. Nanoeng.* **2022**, *8*, 94. [CrossRef]
90. Wu, M.; Ozcelik, A.; Rufo, J.; Wang, Z.; Fang, R.; Jun Huang, T. Acoustofluidic separation of cells and particles. *Microsyst. Nanoeng.* **2019**, *5*, 32. [CrossRef]
91. Fogel, R.; Limson, J.; Seshia, A.A. Acoustic biosensors. *Essays Biochem.* **2016**, *60*, 101–110. [CrossRef] [PubMed]
92. Lakshmanan, A.; Jin, Z.; Nety, S.P.; Sawyer, D.P.; Lee-Gosselin, A.; Malounda, D.; Swift, M.B.; Maresca, D.; Shapiro, M.G. Acoustic biosensors for ultrasound imaging of enzyme activity. *Nat. Chem. Biol.* **2020**, *16*, 988–996. [CrossRef] [PubMed]
93. Barie, N.; Rapp, M. Covalent bound sensing layers on surface acoustic wave (SAW) biosensors. *Biosens. Bioelectron.* **2001**, *16*, 979–987. [CrossRef] [PubMed]
94. Saitakis, M.; Gizeli, E. Acoustic sensors as a biophysical tool for probing cell attachment and cell/surface interactions. *Cell. Mol. Life Sci. CMLS* **2012**, *69*, 357–371. [CrossRef]
95. Mejia Morales, J.; Glynne-Jones, P.; Vassalli, M.; Lippi, G.L. Acoustofluidic interferometric device for rapid single-cell physical phenotyping. *Eur. Biophys. J.* **2022**, *51*, 185–191. [CrossRef] [PubMed]
96. Beabout, K.; Bernhards, C.B.; Thakur, M.; Turner, K.B.; Cole, S.D.; Walper, S.A.; Chavez, J.L.; Lux, M.W. Optimization of Heavy Metal Sensors Based on Transcription Factors and Cell-Free Expression Systems. *ACS Synth. Biol.* **2021**, *10*, 3040–3054. [CrossRef] [PubMed]
97. Gongi, W.; Rube, M.; Ben Ouada, H.; Ben Ouada, H.; Tamarin, O.; Dejous, C. Elaboration and Characterization of a New Heavy Metal Sensor Functionalized by Extracellular Polymeric Substances Isolated from a Tunisian Thermophilic Microalga Strain *Graesiella* sp. *Sensors* **2023**, *23*, 803. [CrossRef] [PubMed]
98. Balashov, S.M.; Rocha, J.M.; Hurtado, M.R.F.; Prestes, J.A.L.; De Campos, A.F.M.; Moshkalev, S.A. Improved Stability and Performance of Surface Acoustic Wave Nanosensors Using a Digital Temperature Compensation. *Front. Sens.* **2021**, *2*, 617484. [CrossRef]
99. Rocha-Gaso, M.I.; March-Iborra, C.; Montoya-Baides, A.; Arnau-Vives, A. Surface generated acoustic wave biosensors for the detection of pathogens: A review. *Sensors* **2009**, *9*, 5740–5769. [CrossRef]
100. Nicolae, I.; Viespe, C.; Miu, D.; Marcu, A. Analyte discrimination by SAW sensor variable loop amplification probing. *Sens. Actuators B Chem.* **2022**, *358*, 131480. [CrossRef]
101. Lange, K.; Rapp, M. Influence of intermediate aminodextran layers on the signal response of surface acoustic wave biosensors. *Anal. Biochem.* **2008**, *377*, 170–175. [CrossRef]
102. Baumgartner, K.; Westerhausen, C. Recent advances of surface acoustic wave-based sensors for noninvasive cell analysis. *Curr. Opin. Biotechnol.* **2023**, *79*, 102879. [CrossRef]
103. Lee, C.-F.; Yan, T.-R.; Wang, T.-H. Long-term monitoring of Caco-2 cell growth process using a QCM-cell system. *Sens. Actuators B Chem.* **2012**, *166–167*, 165–171. [CrossRef]
104. Michl, J.; Park, K.C.; Swietach, P. Evidence-based guidelines for controlling pH in mammalian live-cell culture systems. *Commun. Biol.* **2019**, *2*, 144. [CrossRef]
105. Wang, T.; Green, R.; Guldiken, R.; Mohapatra, S.; Mohapatra, S. Multiple-layer guided surface acoustic wave (SAW)-based pH sensing in longitudinal FiSS-tumoroid cultures. *Biosens. Bioelectron.* **2019**, *124–125*, 244–252. [CrossRef] [PubMed]
106. Baur, J.E.; Spaine, T.W. Electrochemical deposition of iridium (IV) oxide from alkaline solutions of iridium(III) oxide. *J. Electroanal. Chem.* **1998**, *443*, 208–216. [CrossRef]
107. Ayad, M.M.; Salahuddin, N.A.; Alghaysh, M.O.; Issa, R.M. Phosphoric acid and pH sensors based on polyaniline films. *Curr. Appl. Phys.* **2010**, *10*, 235–240. [CrossRef]
108. Hianik, T. Advances in Electrochemical and Acoustic Aptamer-Based Biosensors and Immunosensors in Diagnostics of Leukemia. *Biosensors* **2021**, *11*, 177. [CrossRef] [PubMed]
109. Tigli, O.; Bivona, L.; Berg, P.; Zaghloul, M.E. Fabrication and Characterization of a Surface-Acoustic-Wave Biosensor in CMOS Technology for Cancer Biomarker Detection. *IEEE Trans. Biomed. Circuits. Syst.* **2010**, *4*, 62–73. [CrossRef] [PubMed]
110. Zhang, S.; Bai, H.; Luo, J.; Yang, P.; Cai, J. A recyclable chitosan-based QCM biosensor for sensitive and selective detection of breast cancer cells in real time. *Analyst* **2014**, *139*, 6259–6265. [CrossRef] [PubMed]

111. Hao, H.C.; Chang, H.Y.; Wang, T.P.; Yao, D.J. Detection of cells captured with antigens on shear horizontal surface-acoustic-wave sensors. *J. Lab. Autom.* **2013**, *18*, 69–76. [CrossRef]
112. Wang, T.; Green, R.; Howell, M.; Martinez, T.; Dutta, R.; Mohapatra, S.; Mohapatra, S.S. The design and characterization of a gravitational microfluidic platform for drug sensitivity assay in colorectal perfused tumoroid cultures. *Nanomedicine* **2020**, *30*, 102294. [CrossRef]
113. Dey, N.; Ashour, A.S.; Mohamed, W.S.; Nguyen, N.G. *Acoustic Sensors in Biomedical Applications;* Springer Briefs in Speech Technology; Springer: Cham, Switzerland, 2019.
114. Vellekoop, M.J. Acoustic wave sensors and their technology. *Ultrasonics* **1998**, *36*, 7–14. [CrossRef]
115. Maresca, D.; Lakshmanan, A.; Abedi, M.; Bar-Zion, A.; Farhadi, A.; Lu, G.J.; Szablowski, J.O.; Wu, D.; Yoo, S.; Shapiro, M.G. Biomolecular Ultrasound and Sonogenetics. *Annu. Rev. Chem. Biomol. Eng.* **2018**, *9*, 229–252. [CrossRef] [PubMed]
116. Jathoul, A.P.; Laufer, J.; Ogunlade, O.; Treeby, B.; Cox, B.; Zhang, E.; Johnson, P.; Pizzey, A.R.; Philip, B.; Marafioti, T.; et al. Deep in vivo photoacoustic imaging of mammalian tissues using a tyrosinase-based genetic reporter. *Nat. Photonics* **2015**, *9*, 239–246. [CrossRef]
117. Bi, R.; Dinish, U.S.; Goh, C.C.; Imai, T.; Moothanchery, M.; Li, X.; Kim, J.Y.; Jeon, S.; Pu, Y.; Kim, C.; et al. In Vivo label-free functional photoacoustic monitoring of ischemic reperfusion. *J. Biophotonics* **2019**, *12*, e201800454. [CrossRef] [PubMed]
118. Chen, M.; Duan, X.; Lan, B.; Vu, T.; Zhu, X.; Rong, Q.; Yang, W.; Hoffmann, U.; Zou, J.; Yao, J. High-speed functional photoacoustic microscopy using a water-immersible two-axis torsion-bending scanner. *Photoacoustics* **2021**, *24*, 100309. [CrossRef]

MDPI AG
Grosspeteranlage 5
4052 Basel
Switzerland
Tel.: +41 61 683 77 34

Molecules Editorial Office
E-mail: molecules@mdpi.com
www.mdpi.com/journal/molecules

www.ingramcontent.com/pod-product-compliance
Lightning Source LLC
Chambersburg PA
CBHW062038220926
51850CB00010B/2593
* 9 7 8 3 7 2 5 8 5 8 8 1 1 *